Teach Mather

The companion website for this series can be found at **www.routledge.com/cw/teachnow**. All of the useful web links highlighted in the book can be found here, along with additional resources and activities.

Being taught by a great teacher is one of the genuine privileges of life. Teach Now! *is an exciting new series that opens up the secrets of great teachers and, step by step, helps trainees and new recruits to the profession to build the skills and confidence they need to become first-rate classroom practitioners.*

Written by a highly skilled practitioner, this practical, classroom-focused guide contains all the support you need to become a great Mathematics teacher. Combining a grounded, modern rationale for learning and teaching with highly practical training approaches, the book guides you through the themes of Mathematics teaching and the skills needed to demonstrate learning. *Teach Now! Mathematics* also offers clear, straightforward advice on classroom practice, lesson planning and working in schools.

Teaching and learning, planning, assessment and behaviour management are all covered in detail, with a host of carefully chosen examples used to demonstrate good practice. Every example is rooted in recent experience in the Mathematics classroom. The commonalities of teaching pedagogy across all subjects are discussed, but this book gets to the heart of the unique nature of this subject. From building confidence to developing problem-solving skills and mathematical literacy, this book considers what the keys to success are in learning, and hence teaching, Mathematics.

There are also chapters on dealing with pressure, excelling in observations, finding the right job and succeeding at interview. Throughout the book, there is a great selection of ready-to-use activities, strategies and techniques that will help put you on the fast track to success in the classroom.

With a strong emphasis on sparking students' interest and enthusiasm in Mathematics, this book is your essential guide as you start your exciting and rewarding career as an outstanding Mathematics teacher.

Julia Upton is Head Teacher at Debenham High School, UK. She has taught Mathematics for over twenty years, in a range of schools, and has always been passionate about making Mathematics accessible and engaging for all students. She has been a leading practitioner, has delivered teacher training to both primary and secondary teachers, has worked with subject leaders in Mathematics and, despite being a head teacher, is still equally committed to her own classroom teaching.

Teach Now!

Series editor: Geoff Barton

Being taught by a great teacher is one of the genuine privileges of life. *Teach Now!* is an exciting new series that opens up the secrets of great teachers and, step by step, helps trainees and new recruits to the profession to build the skills and confidence they need to become first-rate classroom practitioners. The series comprises a core text that explores what every teacher needs to know about essential issues, such as learning, pedagogy, assessment and behaviour management, and subject-specific books that guide the reader through the key components and challenges in teaching individual subjects. Written by expert practitioners, the books in this series combine an underpinning philosophy of teaching and learning alongside engaging activities, strategies and techniques to ensure success in the classroom.

Titles in the series:

Teach Now! The Essentials of Teaching
Geoff Barton

Teach Now! History
Becoming a Great History Teacher
Mike Gershon

Teach Now! English
Becoming a Great English Teacher
Alex Quigley

Teach Now! Science
The Joy of Teaching Science
Tom Sherrington

Teach Now! Modern Foreign Languages
Becoming a Great Teacher of Modern Foreign Languages
Sally Allan

Teach Now! Mathematics
Becoming a Great Mathematics Teacher
Julia Upton

Teach Now! Mathematics

Becoming a Great Mathematics Teacher

Julia Upton

LONDON AND NEW YORK

First published 2015
by Routledge
2 Park Square, Milton Park, Abingdon, Oxon OX14 4RN

and by Routledge
711 Third Avenue, New York, NY 10017

Routledge is an imprint of the Taylor & Francis Group, an informa business

British Library Cataloguing in Publication Data
A catalogue record for this book is available from the British Library

Library of Congress Cataloging in Publication Data
Upton, Julia (Julia Rose), author.
Teach now! Mathematics: becoming a great mathematics teacher/Julia Upton.
pages cm
Includes bibliographical references and index.
1. Mathematics – Vocational guidance. 2. Teaching – Vocational guidance. 3. Mathematics teachers – Life skills guides. I. Title.
QA10.5.U68 2015
510.71 – dc23
2014019755

ISBN: 978-1-138-78316-4 (hbk)
ISBN: 978-1-138-78317-1 (pbk)
ISBN: 978-1-315-76881-6 (ebk)

Typeset in Celeste and Optima
by Florence Production Ltd, Stoodleigh, Devon, UK

Contents

Series editor's foreword

What is this series about, and who is it for?

Many of us unashamedly like being teachers.

We shrug off the jibes about being in it for the holidays. We ignore the stereotypes in soap operas, sitcoms, bad films and serious news programmes. We don't feel any need to apologise for what we do, despite a constant and corrosive sense of being undervalued.

We always knew that being criticised was part of the deal.

We aren't defensive. We aren't apologetic. We simply like teaching.

And, whether we still spend the majority of our working week in the classroom or, as senior leaders, we regard the classroom as a sanctuary from the swirling madness beyond the school gates, we think teaching matters.

We think it matters a lot.

And we think that students need more good teachers.

That's where *Teach Now!* started as a concept. Could we – a group of teachers and teaching leaders, scattered across England – put together the kind of books that we wish we had had when we were embarking on our own journeys into the secret garden of education.

Of course, there were lots of books around then. Nowadays, there are even more – books, plus ebooks, blogs and tweets. You can hardly move on the Internet without tripping over another reflection on a lesson that went well or badly, or another teacher extolling a particular approach or dismissing another craze or moaning about the management.

So we know you don't necessarily think you need us. There are plenty of people out there ready to shovel advice and guidance towards a fledgling teacher.

But we wanted to do something different. We wanted to provide two essential texts that would distil our collective knowledge as teachers and package it in a form that was easy to read, authoritative, re-readable, reassuring and deeply rooted in the day-to-day realities of education as it is – not as a consultant or adviser might depict it.

We are writing, in other words, in the early hours of days when each of us will be teaching classes, taking assemblies, watching lessons, looking at schemes of work and dealing with naughty students – and possibly naughty teachers.

We believe this gives our series a distinctive sense of being grounded in the realities of real schools, the kind of places we each work in every day.

We want to provide a warts-and-all account of how to be a great teacher, but we also each believe that education is an essentially optimistic career.

However grim the news out there, in our classrooms, we can weave a kind of magic, given the right conditions and the right behaviour. We can reassure ourselves and the students in front of us that, together, we can make the world better.

And if that seems far-fetched, then you haven't seen enough great teachers.

As Roy Blatchford – himself an exceptional teacher and now the Director of the National Education Trust – says in his list of what great teachers do:

> The best teachers are children at heart
> Sitting in the best lessons, you just don't want to leave.
>
> (Roy Blatchford, *The 2012 Teachers' Standards in the Classroom*, Sage, 2013)

We want young people to experience more lessons like that – classrooms where the sense of time is different, where it expands and shrinks, as the world beyond the classroom recedes, and where interest and passion and fascination take over; places where, whatever your background, your brain will fire with new experiences, thoughts and ideas; where, whatever your experience so far of the adult world, here, in this classroom, is an adult who cares a lot about something, can communicate it vividly and, in the way she or he talks and behaves, demonstrates a care and interest in you that is remarkable.

We need more classrooms like that and more teachers to take their place within them.

So that's what we have set out to do: to create a series of books that will – if you share our sense of moral purpose – help you to become a great teacher.

You'll have noticed that we expect you to buy two books. We said we were optimistic. That's because we think that being a great teacher has two important dimensions to it. First, you need to know your subject – to really know it.

We know from very good sources that the most effective teachers are experts in what they teach. That doesn't mean they know everything about it. In fact, they often fret about how little they feel they truly know. But they are hungry and passionate and eager – and all those other characteristics that define the teachers who inspire us.

So we know that subject knowledge is really important – and not just for teaching older students. It is as important when teaching Year 7s, knowing what you need to teach and what you can, for now, ignore.

We also believe that subject knowledge is much more than a superficial whisk through key dates or key concepts. It's about having a depth of knowledge that allows us to join up ideas, to explore complexity and nuance, to make decisions about what the key building-blocks of learning a subject might be.

Great teachers sense this and, over a number of years, they build their experience and hone their skills. That's why we have developed subject specialist books for English, Mathematics, History, Modern Foreign Languages and Science. These are the books that will help you to take what you learned on your degree course and to think through how to make that knowledge and those skills powerfully effective in the classroom.

They will take you from principles to practice, from philosophy deep into pedagogy. They will help to show you that any terror you may have about becoming a teacher of a subject is inevitable, and that knowing your stuff, careful planning, informed strategies – that all of these will help you to teach now.

Then there's *Teach Now! The Essentials of Teaching*, which is the core text, because we also believe that, even if you are the best-informed scientist, linguist or mathematician in the universe, that this in itself won't make you a great teacher.

That's because great teachers do things that support and supplement their subject knowledge. This is the stuff that the late, great educator Michael Marland called the 'craft of the classroom'. It's what the best teachers know and do instinctively, but, to those of us looking on from the outside, or in the earliest stages of a teaching career, can seem mysterious, unattainable, a kind of magic.

It's also the kind of stuff that conventional training may not sufficiently cover.

We're talking about how to open the classroom door, knowing where to stand, knowing what to say to the student who is cheeky, knowing how to survive when you feel, in the darkest of glooms, intimidated by preparation and by marking, that you've made a terrible career choice.

These two texts combined – the subject specialist book and the core book – are designed to help you wherever you are training – in a school or academy or on a PGCE course. Whether you are receiving expert guidance, or it's proving to be more mixed, we hope our ideas, approaches and advice will reassure you and help you to gain in confidence.

We hope we are providing books that you will want to read and re-read as you train, as you take up your first post and as you finally shrug off the feelings of early insecurity and start to stretch your wings as a fully-fledged teacher.

So that's the idea behind the books.

And throughout the writing of them, we have been very conscious that – just like us – you have too little time. We have therefore aimed to write in a style that is easy to read, reassuring, occasionally provocative and opinionated. We don't want to be bland: teaching is too important for any of us to wilt under a weight of colourless eduspeak.

That's why we have written in short paragraphs, short chapters, added occasional points for reflection and discussion, comments from trainee and veteran teachers, and aimed throughout to create practical, working guides to help you teach now.

So, thanks for choosing to read what we have provided. We would love to hear how your early journey into teaching goes and hope that our series helps you on your way into and through a rewarding and enjoyable career.

Geoff Barton
with Sally Allan, Mike Gershon, Alex Quigley,
Tom Sherrington and Julia Upton
The *Teach Now!* team of authors

Acknowledgements

I would like to express my very great appreciation to all the colleagues whom I have worked with and been lucky enough to encounter in my teaching career. From those first days in teacher training, to local network meetings, in person at national courses and via the web, they have all influenced my practice.

However, I would like to offer my special thanks to all the students whom I have taught who have shaped my thinking, as well as theirs. They are what makes this job so special and they keep me on my toes every day.

Introduction

Like many good lessons, this book starts with an objective: to provide all those learning to teach Mathematics (as a specialist and non-specialist) with realistic, practical advice on how to do a good job.

I hope that, on reading this book, you feel that I have met this aim.

You often hear sports stars and celebrities of screen and stage paying tribute to a teacher who inspired and motivated them when they were younger. You don't often, however, hear scientists or notable academics extolling the wonders that their Mathematics teachers opened their eyes to in their youth. This is not because they didn't have such a person, nor that teachers of these subjects aren't inspirational and motivating in the same way. It is simply that we don't take time to hear the words of those in such professions revealing their starting point and inspiration.

So, you want to enter one of the most rewarding professions, in one of the most challenging subjects. Why would you want to do that? My response would simply be that teaching is a wonderful job, and Mathematics is a fascinating and inspiring subject. Contrary to popular belief and the infamous parental words of, 'I couldn't do Maths at school', there are generations of young people who are eager to learn the fundamentals of numbers and who love the problem-solving, logical, ordered nature of Mathematics. This is a subject that develops rigour, attention to detail and the ability to see complex problems in manageable steps. These reasons are why we must continue to inspire every young person to love Mathematics, whether they need it for their future as a functional tool, or if they use the subject in its purest sense at university and beyond. The skills that learning Mathematics brings are essential and transferable to all walks of life, and this book aims to dispel the myth that it is 'an impossible subject for some'.

This book will take you through your understanding of the variety and mass of content in the Mathematics curriculum (though it will not endeavour to cover all the different topic areas and how to teach them – another book perhaps!) and how you will grapple with this wonderful subject, with specific teaching skills that help to make you the master in the Mathematics classroom.

We will start by gaining an understanding of the essential building-blocks in Mathematics and what this means about teaching the subject. Mathematics teachers come from many backgrounds: although I came from a degree in Mathematics and Computing and went straight into a career in teaching, this is not necessarily the case for all teachers today. Many now come to teaching Mathematics as a second subject, a second career or as an addition to their teaching timetable. Whether you are a proficient mathematician yourself, with a first from Oxbridge, or whether this is a subject that is less familiar to you, I think that we all have something to learn about how we teach the subject. Actually, as a proficient mathematician, I think I had as much to learn as anyone. I can remember vividly the first lesson that I taught, on solving equations, to a Year 9 group. I gave what I thought was a clear, structured and actually perfect explanation of how to use inverse operations to balance an equation. However, when they came to practise the skill, some of them couldn't do it! As a confident mathematician, I just could not understand what there was to not understand. In many ways, those of us with the most expertise have the most to learn as we enter the classroom. Those who come to Mathematics as a second option often have a far greater empathy with students when they do not know what to do and, hence, can be far better at explaining concepts.

It is intended that, whatever your Mathematics background, this book will have something for you to learn. The chapters include real examples used in the classroom that you can try as you are training, questions for you to consider as you develop your own teaching style and activities for you to have a go at, to help you to conceptualise what the inherent skills of Mathematics teaching are and how they might be particular to this subject, as opposed to other subjects.

The book covers curriculum essentials, planning in the short, medium and long term, meaningful assessment in the Mathematics classroom, how to differentiate (even when you teach in sets!), questioning and the one that worries so many at the start of their career, behaviour. This as well as: literacy – do we really have to bother in Mathematics? – and how this is more than just vocabulary; writing that first letter of application; performing at interview and in observations; and surviving that first term in post. It is expected that this book will guide you through your trainee year and those first few years in post. There is no substitute for the experience itself, but there are many things that you can learn before starting in the classroom that will help to make the journey smoother. This book aims to demystify those tricks of the trade and make the impossible seem possible.

We will start by looking at what the Mathematics curriculum looks like. This is not a substitute for an examination textbook, but it sets out what the foundations of learning are in Mathematics, the skills that we, as secondary teachers, hope students will already have when they join us. It will examine how those building-blocks are the foundations for more challenging concepts and how the myriad of different skills link together across the areas of number, algebra, shape and space and data handling. From this starting point and understanding of the curriculum, we will be asking you to consider what teaching in the Mathematics' classroom should look like.

Throughout the book, there will be activities and questions to make you think about what you want learning to look like in your classroom and how this links to what research and the experts say works and doesn't work when students learn Mathematics. I shall start by asking you to consider where you stand. Complete the next page now, and then perhaps again after you have read the book, or one year into your teaching career, to see if your viewpoint has changed.

ACTIVITY: WHERE DO YOU STAND?

Below, for each question, there are two opposing statements, which represent each end of the spectrum in a particular area of Mathematics teaching. Using a scale of 1 to 10 (with 1 and 10 referring to the extreme statements given) where would you sit?

1 What do I think learning looks like?

1 – 2 – 3 – 4 – 5 – 6 – 7 – 8 – 9 – 10

Student-led learning: pupils should learn from each other; the teacher should act as a facilitator.

Teacher-led learning: learning should be delivered and led by the teacher.

2 What am I responsible for?

1 – 2 – 3 – 4 – 5 – 6 – 7 – 8 – 9 – 10

Responsible for education: the Maths teacher's sole responsibility should be the attainment and achievement of the pupils.

Responsible for the whole child: the Maths teacher is part of a pupil's world and so must consider this when teaching within the classroom.

3 Why do we teach Mathematics?

1 – 2 – 3 – 4 – 5 – 6 – 7 – 8 – 9 – 10

For its own sake: the Mathematics curriculum should pave the way for the explicit study of Mathematics as a subject in its own right.

For others: Mathematics is a tool that supports exploration of information in other subject areas and the world at large.

4 My classroom will have . . .

1 – 2 – 3 – 4 – 5 – 6 – 7 – 8 – 9 – 10

A peaceful silence: Maths lessons should be quiet, scholarly, calm places, like the Reading Room of the British Library.

A buzz of activity: Maths lessons should be humming with activity, like the bustling hall in the Natural History Museum.

5 What do I think about setting in Mathematics?

1 – 2 – 3 – 4 – 5 – 6 – 7 – 8 – 9 – 10

Integration: the best Maths teaching happens when students are taught in mixed-ability groups, raising the aspirations of all learners.

Segregation: Mathematics is best taught in rigid sets, where the ability of the student alone places them in a group.

This book aims to help you to answer these questions and more, as you develop what you understand to be outstanding Mathematics teaching, what it feels like in your classroom, and what you explicitly do to make this happen. This book will give some examples of research and modern mathematical thinking, but it is not a book with hollow statements and generic suggestions. It gives specific examples that can be tried in the classroom and that have been used by others before you!

Enjoy.

1 Curriculum essentials

This chapter will look at what is contained within the Mathematics curriculum. What does the National Curriculum say, what is in a typical GCSE specification, and what do these both imply for the teaching of Mathematics? It will look at the breadth of topics within the Mathematics curriculum, mainly at Key Stages 3 and 4, and consider how this very content-heavy subject can become more manageable in its delivery and improve student outcomes. It will consider the standing of the UK in world league tables for Mathematics education and the implications of this for the UK.

What is contained within the Mathematics curriculum?

This is, of course, a question that needs a whole book to be answered, and I will not attempt to cover all aspects of the National Curriculum and what it means you will have to teach in one chapter. What I do want readers of this book to understand from this chapter is the breadth and range of topics that are covered, and what this will imply for you as a teacher. In your use of Mathematics as a degree student, or in the workplace, there will be many things that you will need to teach that you have not utilised for a very long time. It is essential that, as you start out in your teaching career, you understand what there is to cover and start to identify any gaps that you might have in your subject knowledge.

It will seem an odd to thing to do, but I am going to start the chapter with a quick set of ten questions, of a type that has a single-step calculation needed for an answer and that you might use as a recap at the start of a lesson. The idea of this is not to set your heart racing but to give you a glimpse, right from the start, of the breadth of the Mathematics curriculum.

I have used these ten questions with groups of teachers who are not mathematicians, and it is incredible to notice what a sense of fear and dread fills the room as soon as you mention that they are going to do some maths. Each question is displayed for at most 30 seconds, and participants must answer at speed. Giving yourself at most 2 minutes for all ten . . . go!

ACTIVITY: THE FEAR FACTOR

1. 16×25
2. $\frac{2}{5}$ of 45
3. Solve $2x - 5 = 14$.
4. 3, 4, 5, 8. What is the mean of these four numbers?
5. Simplify $a^2 \times a^3$.
6. Increase 60 by 20%.
7. 0.3×0.2
8. Simplify $\frac{12}{18}$.
9. $-2 - -5$
10. Calculate angle b in the figure.

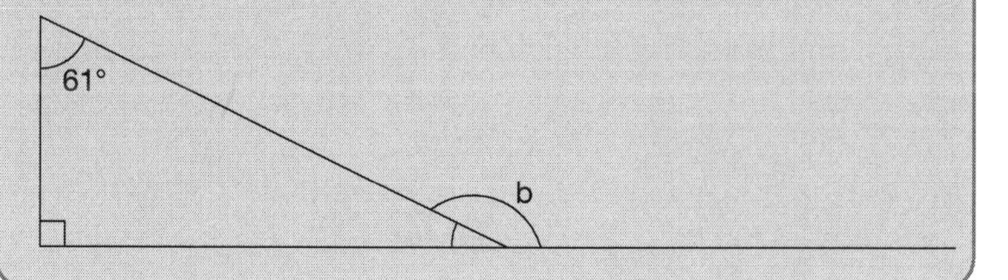

How did you fare?

I would hope that you found that set of questions fairly simple, but what does it tell us about the Mathematics curriculum? Each of the ten questions could be simple questions at the lower end of GCSE or the earlier stages of the secondary curriculum. However, what I want to demonstrate, despite the apparent simplicity of these ten questions, is what each might imply for you as a teacher.

Why would this list of fairly straightforward sums be so daunting to so many non-mathematicians? Indeed, is this what makes the subject so scary to so many young people? Let us take each question in turn and consider a few things.

(1) 16×25

How did you work this out? Was it $16 \times 100 \div 4$? Or did you use a formal column method, or perhaps a Chinese grid method. As competent mathematicians, we approach this question in a way that suits the numbers involved. However, in doing so, we are using a sophisticated understanding of multiplication and are choosing our problem-solving method accordingly.

(2) $\frac{2}{5}$ of 45

The vast majority of people (young and old alike) would be far less daunted if this question were merely one-fifth of 45. Although many can see how one-fifth requires the total to be split into five, when two-fifths is added to the complexity, this makes them perform all sorts of weird and wonderful calculations.

(3) Solve $2x - 5 = 14$.

We will tackle in Chapter 6 the question of language in Mathematics. As those skilled in its art, we know the difference between a question that says 'solve' and one that says 'simplify' or 'evaluate'. These words are key triggers to what you then need to do. We need to make these explicit to our students.

(4) 3, 4, 5, 8. What is the mean of these four numbers?

Students are generally very happy with the concepts of mean, median and mode, although still see the mean as the one that is the average! This question is generally well done, but is a reminder that language is important and worth considering. Also, how do students fare on a tabular presentation of data, when asked to calculate the mean, if they have learned that it is 'add them all up and divide by how many there are'?

(5) Simplify $a^2 \times a^3$.

Not too complex, but a reminder of the many rules and protocols that are required for Mathematics success. These can be learned by rote, but are far better learned with an understanding of why.

(6) Decrease 60 by 20%.

The calculation may not be hard, but many leave it at 12 and forget the decrease part. As a next step, why would you want students to learn this by doing 60 × 0.8 on their calculator, rather than using a mental method? This is one example of where they cling to a method learned earlier, when things become more complex, only to find that their method has its limitations.

(7) 0.3 × 0.2

My bet is that, although almost all could do 3 × 2, many cannot do this. How do we build that concept of place value in all that we do?

(8) Simplify $^{12}/_{18}$.

It might be that students recognise the highest common factor and cancel by a factor of 6. However, it is most likely that they will divide by two to get $^{6}/_{9}$, but many will not then progress to $^{2}/_{3}$. How do you get students to consider their answer and ensure it is as accurate and correct as it can be, even after they have performed a calculation that, in essence, is not incorrect?

(9) $-2 - -5$

The Achilles heel of even some exceptional A Level mathematicians – negative numbers. To be treated with due care and respect at all times!

(10) Calculate angle b.

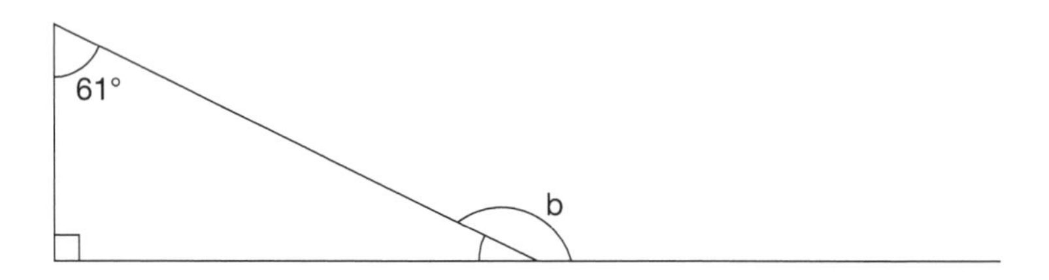

Finally, a reminder that Mathematics is about shape and geometry, as well as number and algebra.

So, from just ten short questions of simple Mathematics, I hope that you can see just how complex this subject can be and how we must carefully consider how we teach it to dispel that oft-heard cry of 'I can't do Maths'.

The breadth of the topics within the Mathematics National Curriculum

One of the greatest challenges within Mathematics is the mountain of individual things that there are to learn. Let's just stop and think a moment about how many different topics a GCSE student needs to cover:

A angles

B bearings

C cosine rule

D diameter

E expansion of brackets

F fractions

G geometric series

H hypotenuse

I indices

J Julia sets (I had to stretch beyond the GCSE syllabus for this letter, I am afraid, and I just couldn't resist this one.)

K kites

L lowest common multiple

M mean (of simple, tabulated and grouped data)

N *n*th terms

O one-to-one functions

P primes

Q quadratic graphs

R reciprocal

S simultaneous equations

T transformations

U upper bounds

V vectors

W whole numbers

X x–y plane

Y y-intercept

Z zero division.

My A–Z list is obviously not all-encompassing and merely gives a flavour of the range and depth of topics that are studied. In delving into this plethora of topics, we must seek to create mathematicians who have mathematical fluency, or, as Lynne McClure, the director of NRICH,[1] put it at a recent National Centre for Excellence in the Teaching of Mathematics (NCETM) event, 'mathematical fluency is the ability to use Mathematics efficiently, accurately and flexibly'.

The aims of Mathematics teaching

The new National Curriculum[2] for study in 2014 is quite right in its aims for the teaching and learning of Mathematics:

> The National Curriculum for mathematics aims to ensure that all pupils:
>
> - become fluent in the fundamentals of mathematics, including through varied and frequent practice with increasingly complex problems over time, so that pupils develop conceptual understanding and the ability to recall and apply knowledge rapidly and accurately
> - reason mathematically by following a line of enquiry, conjecturing relationships and generalisations, and developing an argument, justification or proof using mathematical language
> - can solve problems by applying their mathematics to a variety of routine and non-routine problems with increasing sophistication, including breaking down problems into a series of simpler steps and persevering in seeking solutions.

So, what does our fear factor test and the aims from the National Curriculum tell us about teaching Mathematics?

We have to help students to move between topics and develop transferable skills that allow them to be proficient in their use of mathematical expertise. As confident mathematicians, we do this without thinking, using knowledge from one area of the curriculum and applying to another. For students within the subject, who have a tendency to compartmentalise what we teach them, we must make this more obvious. Let us take one question and see what assumptions we might make from it.

ACTIVITY: WHAT SKILLS OR KNOWLEDGE ARE NEEDED TO COMPLETE THIS QUESTION?

The cost of producing labels for cans is 2.1 p/cm^2. Calculate the cost of producing the label for this tin of baked beans.

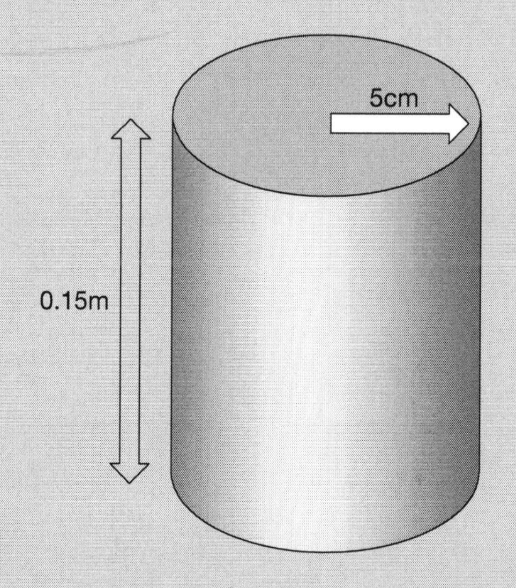

Jot down here any thoughts that you have about what this question involves.

Circumference of a circle

Conversion

Area (circ x 0.15m)

What are the skills needed to be able to answer this question?

- *Surface area*: recognition that this question needs surface area and not volume is something some students will miss.
- *Circumference of a circle*: the knowledge that the circumference of the circle on the base of the cylinder forms the length of the rectangle that makes the sides of cylinder is one that the students struggle to understand – visual examples of a rolling wheel, a garden roller and an unfurled label off a can of soup are all helpful ways of visualising this.

These two above of course expect recognition of the correct rule and the recall of those rules, but what else?

- *Units*: recognition that the can has two lengths in different units and how to deal with these. Most students will convert happily between centimetres and metres, but forethought in this question as to which would be best is helpful. If the problem is taken in metres in the first instance, then, when the time comes to use the 2.1p/cm^2 value, the conversion from metres squared to centimetres squared is not as obvious. It can take quite some explaining to show students why 100 cm = 1 m does not imply that 100 cm^2 = 1 m^2. Of course, the question was also not explicit about the units desired in the final answer, which we may well realise, after calculation, would be better in pounds rather than pence.
- *Use of a 'per' calculation*: this can need some thought. Students will often dive into a multiplication calculation, when in fact division may be needed. What is the unit saying, and what does that imply for the information that you have? A confident grasp of the four operations is essential here.

To answer questions such as this, we must, from the word go, help students to build links between topics all the time. We might be teaching how to work out the circumference of a circle, but we are also, hence, teaching the use of a formula, the use of units and approximations, estimating solutions and accuracy of answers. It is easy, when you start teaching Mathematics, to compartmentalise each topic area and mentally 'tick it off' when you have covered it. It will take time for you to develop a working understanding of the curriculum and all the topics within it. However, from the start, you can be considering what links there are between topics and how each new skill builds on previous foundations.

Let's look in more detail at the National Curriculum statements and then at another topic area and how it could break down.

The National Curriculum in more detail

The National Curriculum gives three broad domains that should be developed through the teaching of mathematical content. They are:

Develop fluency

- consolidate their numerical and mathematical capability from Key Stage 2 and extend their understanding of the number system and place value to include decimals, fractions, powers and roots
- select and use appropriate calculation strategies to solve increasingly complex problems
- use algebra to generalise the structure of arithmetic, including to formulate mathematical relationships
- substitute values in expressions, rearrange and simplify expressions, and solve equations
- move freely between different numerical, algebraic, graphical and diagrammatic representations [for example, equivalent fractions, fractions and decimals, and equations and graphs]
- develop algebraic and graphical fluency, including understanding linear and simple quadratic functions

- use language and properties precisely to analyse numbers, algebraic expressions, 2D and 3D shapes, probability and statistics

Reason mathematically

- extend their understanding of the number system; make connections between number relationships, and their algebraic and graphical representations
- extend and formalise their knowledge of ratio and proportion in working with measures and geometry, and in formulating proportional relations algebraically
- identify variables and express relations between variables algebraically and graphically
- make and test conjectures about patterns and relationships; look for proofs or counter-examples
- begin to reason deductively in geometry, number and algebra, including using geometrical constructions
- interpret when the structure of a numerical problem requires additive, multiplicative or proportional reasoning
- explore what can and cannot be inferred in statistical and probabilistic settings, and begin to express their arguments formally

Solve problems

- develop their mathematical knowledge, in part through solving problems and evaluating the outcomes, including multi-step problems
- develop their use of formal mathematical knowledge to interpret and solve problems, including in financial mathematics
- begin to model situations mathematically and express the results using a range of formal mathematical representations

- select appropriate concepts, methods and techniques to apply to unfamiliar and non-routine problems

The National Curriculum is then broken into different topic areas at Key Stage 3. These are:

- algebra;
- ratio, proportion and rates of change;
- number;
- geometry and measure;
- probability;
- statistics.

On the next few pages are the brief topic areas to be covered at Key Stage 3; they of course assume that students have mastered and have at their fingertips all of the preceding topics at Key Stage 2!

ACTIVITY: THE KEY STAGE 3 NATIONAL CURRICULUM

Although I don't want to make you panic, I do want you to see, from the start, the range of topics, even at the start of the secondary stage. Take a look at these topic headings and, with a highlighter, indicate your confidence in how you might teach each topic area. Think about how you would teach it, rather than just whether you know how to do it.

- Green highlighter: I know this topic well and can think of some activities that I might use to teach it.
- Amber highlighter: I know this topic well but would struggle to think about how I might teach it.
- Red highlighter: I can't remember much about this topic and would need to review it carefully before teaching it.

Algebra

Pupils should be taught to:

- use and interpret algebraic notation, including:
  - ab in place of $a \times b$
  - $3y$ in place of $y + y + y$ and $3 \times y$
  - a^2 in place of $a \times a$, a^3 in place of $a \times a \times a$; a^2b in place of $a \times a \times b$
  - a/b in place of $a \div b$
  - coefficients written as fractions rather than as decimals
  - brackets
- substitute numerical values into formulae and expressions, including scientific formulae
- understand and use the concepts and vocabulary of expressions, equations, inequalities, terms and factors
- simplify and manipulate algebraic expressions to maintain equivalence by:
  - collecting like terms
  - multiplying a single term over a bracket
  - taking out common factors
  - expanding products of 2 or more binomials
- understand and use standard mathematical formulae; rearrange formulae to change the subject
- model situations or procedures by translating them into algebraic expressions or formulae and by using graphs
- use algebraic methods to solve linear equations in 1 variable (including all forms that require rearrangement)
- work with coordinates in all 4 quadrants
- recognise, sketch and produce graphs of linear and quadratic functions of 1 variable with appropriate

scaling, using equations in x and y and the Cartesian plane

- interpret mathematical relationships both algebraically and graphically
- reduce a given linear equation in two variables to the standard form $y = mx + c$; calculate and interpret gradients and intercepts of graphs of such linear equations numerically, graphically and algebraically
- use linear and quadratic graphs to estimate values of y for given values of x and vice versa and to find approximate solutions of simultaneous linear equations
- find approximate solutions to contextual problems from given graphs of a variety of functions, including piece-wise linear, exponential and reciprocal graphs
- generate terms of a sequence from either a term-to-term or a position-to-term rule
- recognise arithmetic sequences and find the nth term
- recognise geometric sequences and appreciate other sequences that arise

Ratio, proportion and rates of change

Pupils should be taught to:

- change freely between related standard units [for example time, length, area, volume/capacity, mass]
- use scale factors, scale diagrams and maps
- express 1 quantity as a fraction of another, where the fraction is less than 1 and greater than 1
- use ratio notation, including reduction to simplest form
- divide a given quantity into 2 parts in a given part:part or part:whole ratio; express the division of a quantity into 2 parts as a ratio

- understand that a multiplicative relationship between 2 quantities can be expressed as a ratio or a fraction
- relate the language of ratios and the associated calculations to the arithmetic of fractions and to linear functions
- solve problems involving percentage change, including: percentage increase, decrease and original value problems and simple interest in financial mathematics
- solve problems involving direct and inverse proportion, including graphical and algebraic representations
- use compound units such as speed, unit pricing and density to solve problems

Geometry and measures

Pupils should be taught to:

- derive and apply formulae to calculate and solve problems involving: perimeter and area of triangles, parallelograms, trapezia, volume of cuboids (including cubes) and other prisms (including cylinders)
- calculate and solve problems involving: perimeters of 2D shapes (including circles), areas of circles and composite shapes
- draw and measure line segments and angles in geometric figures, including interpreting scale drawings
- derive and use the standard ruler and compass constructions (perpendicular bisector of a line segment, constructing a perpendicular to a given line from/at a given point, bisecting a given angle);

recognise and use the perpendicular distance from a point to a line as the shortest distance to the line
- describe, sketch and draw using conventional terms and notations: points, lines, parallel lines, perpendicular lines, right angles, regular polygons, and other polygons that are reflectively and rotationally symmetric
- use the standard conventions for labelling the sides and angles of triangle ABC, and know and use the criteria for congruence of triangles
- derive and illustrate properties of triangles, quadrilaterals, circles, and other plane figures [for example, equal lengths and angles] using appropriate language and technologies
- identify properties of, and describe the results of, translations, rotations and reflections applied to given figures
- identify and construct congruent triangles, and construct similar shapes by enlargement, with and without coordinate grids
- apply the properties of angles at a point, angles at a point on a straight line, vertically opposite angles
- understand and use the relationship between parallel lines and alternate and corresponding angles
- derive and use the sum of angles in a triangle and use it to deduce the angle sum in any polygon, and to derive properties of regular polygons
- apply angle facts, triangle congruence, similarity and properties of quadrilaterals to derive results about angles and sides, including Pythagoras' Theorem, and use known results to obtain simple proofs

- use Pythagoras' Theorem and trigonometric ratios in similar triangles to solve problems involving right-angled triangles
- use the properties of faces, surfaces, edges and vertices of cubes, cuboids, prisms, cylinders, pyramids, cones and spheres to solve problems in 3D
- interpret mathematical relationships both algebraically and geometrically

Probability

Pupils should be taught to:

- record, describe and analyse the frequency of outcomes of simple probability experiments involving randomness, fairness, equally and unequally likely outcomes, using appropriate language and the 0–1 probability scale
- understand that the probabilities of all possible outcomes sum to 1
- enumerate sets and unions/intersections of sets systematically, using tables, grids and Venn diagrams
- generate theoretical sample spaces for single and combined events with equally likely, mutually exclusive outcomes and use these to calculate theoretical probabilities

Statistics

Pupils should be taught to:

- describe, interpret and compare observed distributions of a single variable through: appropriate graphical representation involving discrete, continuous and grouped data; and appropriate measures of central tendency (mean, mode, median) and spread (range, consideration of outliers)

- construct and interpret appropriate tables, charts, and diagrams, including frequency tables, bar charts, pie charts, and pictograms for categorical data, and vertical line (or bar) charts for ungrouped and grouped numerical data
- describe simple mathematical relationships between 2 variables (bivariate data) in observational and experimental contexts and illustrate using scatter graphs

I hope that, by looking at the Key Stage 3 National Curriculum areas and considering some questions, you have started to realise the contents and challenges to the curriculum in Mathematics. As a next step, it would be worth looking at some examples of GCSE assessment to assess your knowledge (in fact, I know that you may have had to sit a GCSE paper as part of your entry on to a training course). Past GCSE exam papers would be a good starting point, but, alongside this, it would be worth taking a look at a GCSE text book or a web-based resource to see how topics are explained and the examples that are used.

As you look at each topic area, consider these three questions:

1 What prior knowledge would you need to be able to do this topic/question?

2 What skills would I need to teach to access this question?

3 How do skills in this question link to other mathematical skills (mathematical fluency)?

As we tackle planning in Chapter 4, we will come back to these ideas.

What is the international picture for Mathematics, and how will this affect my teaching?

With so much in the press about the standing of the UK with regard to Mathematics and education in general, I want to pause for a moment to consider what has been said about Mathematics education and the implications for the UK classroom.

There has been much concern expressed over the past few years regarding the standards of mathematical instruction in schools. Critical voices have been heard from both employers of young people and from tutors in higher education, where some degree courses have been lengthened to allow time for remedial work in Mathematics. Results from international comparative studies have been depressing news for British Mathematics teaching. Despite the fact that GCSE results in Mathematics have improved (although many would doubt the validity of this), we are still told that there is a shortage of young people pursuing this subject further and a lack of numerical competence in people entering the workplace.

In my experience, the numbers of students progressing to A level Mathematics increased from around 2000 and continues to grow today. However, the numbers of students progressing to degree-level Mathematics are still sparse. Many are taking scientific subjects at degree level and utilising Mathematics as a tool for these studies, but we still lack students entering fields such as engineering, which uses so many mathematical skills and concepts. At a recent careers fair in my own school, the number of students who signed up for workshops on art and design far outweighed the numbers who wanted to attend the talks by those from the energy sector and civil engineering.

Jo Boaler, author of *The Elephant in the Classroom*,[3] said:

> Far too many students hate maths. As a result adults all over the world fear maths and avoid it at all costs. . . . It is the subject that above all else can make them feel helpless and stupid . . .

> it has the power to crush all their confidence and deter them from learning for years to come.

We are constantly reminded of OECD (Organisation for Economic Co-operation and Development) reports and PISA (Programme for International Student Assessment) studies that place the UK at the bottom of the league tables, for education, but specifically for Mathematics. In February 2014, the latest OECD report compared not only students from one country with those from another but those from different socio-economic classes within that country. There has been much talk about social mobility and the evidence that, in the UK, class does matter.

The Spirit Level: Why equality is better for everyone[4] shows, through a number of studies, how the more equal a society is in financial terms, the better off we all are in all life's important measures: physical health, mortality, education, mental health, imprisonment, obesity, trust, community life, violence, crime, teenage pregnancies and more. The evidence presented by Wilkinson and Pickett suggests that those in higher and lower social classes benefit from this greater equilibrium. How that financial equilibrium is achieved is not of importance; in the Scandinavian countries at the top of the league tables it is by high taxes for those earning greater incomes; in the eastern countries such as China it is managed by a limit on the salaries of those at the top of organisations. Interestingly, in November 2013, Switzerland voted no to a similar structure. Swiss voters rejected a proposal that would have limited executive pay to twelve times that of the lowest paid. From a country that is home to a range of giant businesses, including pharmaceutical companies Novartis and Roche, the insurance groups Zurich and Swiss Re and the banks UBS and Credit Suisse, this result is perhaps not a surprise.

The OECD survey in February 2013 separated those from different backgrounds. Elizabeth Truss, the education minister, said that English schools need to 'adopt the teaching practices and positive philosophy' that characterise schools in parts of the Far

East: 'They have a can-do attitude to maths, which contrasts with the long-term anti-maths culture that exists here'.

The survey goes on to show how the children of UK professionals scored an average of 526 on the assessment. This was overshadowed by 656 scored by the children of professionals in Shanghai, China, and 569 from children of low-grade workers in the same region. The children of equivalent parents in unskilled jobs in the UK scored an average of 461.

What the government ministers and those from education who accompany them on their fact-finding mission will discover is that the societal structure, work ethic across all employees and desire to do better than their forefathers are all inextricably linked to why the Far East outperforms the UK. However, they will also find that their manner of teaching Mathematics also has some very particular structures, and that classes in Shanghai are very different to those in the UK. These differences and the implications for success for us will be explored further in our pedagogy focus in Chapter 2.

It is clear, as GCSE reform takes place as I write, that in the course of the next few years our curriculum and the assessment of that curriculum in the UK will be influenced by the practice in the perceived 'best' nations in the world rankings.

You have had a go at my ten quick starter questions. Let us now have a go at a sample of OECD Mathematics questions.

ACTIVITY: OECD SAMPLE MATHEMATICS QUESTIONS

1 Helen has just got a new bike. It has a speedometer that sits on the handlebar. The speedometer can tell Helen the distance she travels and her average speed for a trip. On one trip, Helen rode 4 km in the first 10 minutes and then 2 km in the next 5 minutes. Which one of the following statements is correct?

(a) Helen's average speed was greater in the first 10 minutes than in the next 5 minutes.

(b) Helen's average speed was the same in the first 10 minutes and in the next 5 minutes.
(c) Helen's average speed was less in the first 10 minutes than in the next 5 minutes.
(d) It is not possible to tell anything about Helen's average speed from the information given.

2 Helen rode 6 km to her aunt's house. Her speedometer showed that she had averaged 18 km/h for the whole trip. Which one of the following statements is correct?

(a) It took Helen 20 minutes to get to her aunt's house.
(b) It took Helen 30 minutes to get to her aunt's house.
(c) It took Helen 3 hours to get to her aunt's house.
(d) It is not possible to tell how long it took Helen to get to her aunt's house.

3 Helen rode her bike from home to the river, which is 4 km away. It took her 9 minutes. She rode home using a shorter route of 3 km. This only took her 6 minutes. What was Helen's average speed, in km/h, for the trip to the river and back?

(a) 28
(b) 28.3
(c) 28.6
(d) 28.9

4 Mount Fuji is a famous dormant volcano in Japan. It is only open to the public for climbing from 1 July to 27 August each year. About 200,000 people climb Mount Fuji during this time. On average, about how many people climb Mount Fuji each day?

(a) 340
(b) 710
(c) 3,400
(d) 7,100

5 The Gotemba walking trail up Mount Fuji is about 9 km long. Walkers need to return from the 18 km walk by 8 p.m. Toshi estimates that he can walk up the mountain at 1.5 km/h on average, and down at twice that speed. These speeds take into account meal breaks and rest times. Using Toshi's estimated speeds, what is the latest time he can begin his walk, so that he can return by 8 p.m.?

(a) 10 a.m.
(b) 11 a.m.
(c) 12 p.m.
(d) 1 p.m.

6 Toshi wore a pedometer to count his steps on his walk along the Gotemba trail. His pedometer showed that he walked 22,500 steps on the way up. Estimate Toshi's average step length for his walk up the 9-km Gotemba trail. Give your answer in centimetres (cm).

(a) 0.4
(b) 0.45
(c) 40
(d) 45

7 A revolving door includes three wings that rotate within a circular-shaped space. The door makes four complete rotations in a minute. There is room for a maximum of two people in each of the three door sectors. What is the maximum number of people that can enter the building through the door in 30 minutes?

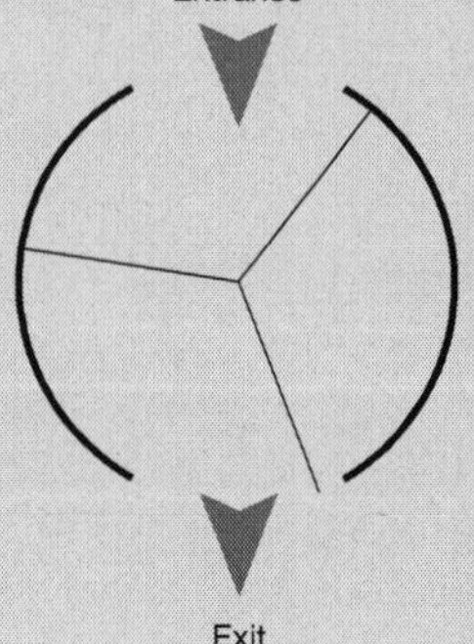

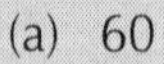

(a) 60
(b) 180
(c) 240
(d) 720

8 Chris has just received her car driving licence and wants to buy her first car. This table shows the details of four cars she finds at a local car dealer. Chris wants a car that meets all of these conditions: the distance travelled is not higher than 120,000 km, it was made in the year 2000 or a later year, and the advertised price is not higher than 4,500 zeds. Which car meets Chris's conditions?

Model:	Alpha	Bolte	Castel	Dezal
Year	2003	2000	2001	1999
Advertised price (zeds)	4,800	4,450	4,250	3,990
Distance travelled (kilometres)	105,000	115,000	128,000	109,000
Engine capacity (litres)	1.79	1.796	1.82	1.783

(a) Alpha
(b) Bolte
(c) Castel
(d) Dezal

9 Which car's engine capacity is the smallest?

Model:	Alpha	Bolte	Castel	Dezal
Year	2003	2000	2001	1999
Advertised price (zeds)	4,800	4,450	4,250	3,990
Distance travelled (kilometres)	105,000	115,000	128,000	109,000
Engine capacity (litres)	1.79	1.796	1.82	1.783

(a) Alpha
(b) Bolte
(c) Castel
(d) Dezal

10 Chris will have to pay an extra 2.5 per cent of the advertised cost of the car as taxes. How much are the extra taxes for the Alpha?

Model:	Alpha	Bolte	Castel	Dezal
Year	2003	2000	2001	1999
Advertised price (zeds)	4,800	4,450	4,250	3,990
Distance travelled (kilometres)	105,000	115,000	128,000	109,000
Engine capacity (litres)	1.79	1.796	1.82	1.783

(a) 120
(b) 160
(c) 1,200
(d) 1,600

Although these are obviously harder than my initial questions, there is clearly no difference in curriculum content across the nations. In Chapter 2, we will consider teaching methodologies, how they differ and what is the best tack for you to take in your classroom to help students achieve their best, but, inherently, the curriculum is the same from one country to another. Mathematics is rare as a subject in this regard, and this is why it is such a common language across the world.

Where does this leave us?

John Allen Paulos, an American professor of Mathematics, writes in a similar vein of the importance of Mathematics to all areas of society, in *Innumeracy, Mathematical Illiteracy and its Consequences.*[5] In this short book, Paulos argues that our inability to deal rationally with very large numbers and the probabilities associated with them results in misinformed governmental policies. That is why we must be cautious about the data and spurious statistics that are coming at us from the PISA and OECD reports at present. There are many factors that affect the results gained in Shanghai and they aren't all related to the type or quality of Mathematics teaching. The one thing that Paulos is certain about is that we must be judicious and competent in our use of statistics and numerical information, if we are to interpret them correctly.

Whatever our views on the standards in our Mathematics education provision, instead of continually criticising our current teaching force and comparing ourselves with those countries that outperform us, we must now invest in and value our teachers, while at the same time encouraging them to increase their expectations for the mathematical attainment of their pupils; support, rather than blame, is needed if we are to rescue British Mathematics education.

At the heart of this text is an appeal to all concerned to work together, cooperatively and constructively, to improve the situation. We need to collaborate to support those entering the profession and work together to enthuse more young people with a love of numbers, logic and reasoning.

ACTIVITIES

- Consider trying an exam paper and highlight which areas you need to review, first, for your own knowledge and, second, to consider how you would teach them.
- Gather some resources to allow you to see different ways that others explain different topics – a GCSE revision guide, a website for students, some text books with worked examples.
- Take one topic area that you haven't covered for some time and consider how you might explain that to students. Would that explanation be different for more able students? For the less able? Would you show

alternative methods? Or just one method? How could you build a conceptual understanding, rather than just teaching a system to follow?

TALKING POINTS

- Have you reviewed the content of GCSE Mathematics? Look at one of the GCSE examination-board websites and download the GCSE higher tier specification. What is covered, and what isn't?
- What areas of Mathematics do you need to review to ensure your confidence in all areas of the curriculum?

Your thoughts

Notes

1 The NRICH Maths Project provides Mathematics resources for children, parents and teachers to enrich learning. See: http://nrich.maths.org/frontpage

2 See: www.gov.uk/government/publications/national-curriculum-in-england-mathematics-programmes-of-study/national-curriculum-in-england-mathematics-programmes-of-study

3 J. Boaler (2010) *The Elephant in the Classroom: Helping children learn and love maths*. London: Souvenir Press.

4 R. G. Wilkinson and K. Pickett (2010) *The Spirit Level: Why equality is better for everyone*. London: Penguin.

5 J. A. Paulos (1988) *Innumeracy: Mathematical illiteracy and its consequences*. New York: Hill & Wang.

2 Pedagogy essentials

This chapter will examine the pedagogy of Mathematics teaching. This will all hinge upon the key to Mathematics learning, confidence building and the ability to break down problems. It will consider classroom discussion as a strategy to build sound reasoning skills versus the importance of consolidation and practice to embed understanding. It will look at the layers of learning, through consideration of Maslow's hierarchy of needs in a teaching and learning context, and the implications of this for your classroom practice.

The big debate in my mind for Mathematics teaching is whether the most effective Mathematics teaching is about exploration and discovery *or* is it about rote learning and consolidation?

I realise that my choice of words for the latter will sound inflammatory in the first instance and also conjure up an image of classrooms with a chalkboard, wooden desks in single file and silent, obedient students. However, it is meant purely to provoke debate and consideration in your own mind about the nature of Mathematics learning.

I will take the first side of the coin for further expansion.

Is the most effective teaching about exploration and discovery?

The Mathematics classroom has changed over the last 20 years, and there are now many good sources to inspire with regard to Mathematics pedagogy. Mathematics is often falsely accused of being a subject where students are told what to do and then to press repeat a hundred times. Now, most educationalists advocate an approach where students explore problems and try to find their own solutions.

In *The Elephant in the Classroom: Helping children learn and love maths,* Jo Boaler[1] quotes a Year 9 student in America talking about her mathematical experience: 'With math you have to interact with everybody and talk to them and answer their questions. You can't be just like "oh here is the book, look at the numbers and figure it out"'.

It is now firmly established that the most effective way to learn is through discovery and doing something for oneself. Much research, originally stemming from Bloom's taxonomy[2] (Table 2.1), has firmly established the place of project work and investigation in the Mathematics classroom.

Bloom's taxonomy in the Mathematics classroom

So how does Bloom's taxonomy apply in the Mathematics classroom? In its most simplistic form, imagine entering a classroom of Year 7 students who have been exploring number patterns and asking them what they are doing. How about these responses, in a Bloom style?

- *Know*: I am continuing the pattern in the sequence, see, by adding 7 each time to get to the next number.
- *Understand*: I am continuing the pattern in the sequence; I have to work out the pattern and then continue it.

Table 2.1 Bloom's taxonomy

Create	Compile information together in a different way by combining elements in a new pattern or proposing alternative solutions. Production of a unique communication
Evaluate	Present and defend opinions by making judgements about information, validity of ideas or quality of work based on a set of criteria judgements in terms of internal evidence. Judgements in terms of external criteria
Analyse	Examine and break information into parts by identifying motives or causes. Make inferences and find evidence to support generalisations
Apply	Use new knowledge. Solve problems in new situations by applying acquired knowledge, facts, techniques and rules in a different way
Understand	Demonstrate understanding of facts and ideas by organising, comparing, translating, interpreting, giving descriptions and stating main ideas
Know	Exhibit memory of previously learned materials by recalling facts, terms, basic concepts and answers

- *Apply*: I am looking at new patterns and seeing if I can work out the pattern from previous examples. I can see from previous problems how this new example might work.
- *Analyse*: I am looking at the patterns and thinking about how to generalise the pattern and know that it will always work.
- *Evaluate*: I am looking at my answers and considering if they are correct and testing them with different numbers to try my theory.
- *Create*: I am making my own number patterns and sequences for others to analyse.

There is no doubt that, in a classroom where topics are explored and developed, students will develop a greater understanding and learn more deeply. But what does this mean for your organisation and culture? The greatest challenge in a problem-solving approach in the classroom is that students will not all reach the same answer at the first time of trying, nor should they, and their answers will not be the same. So, the classroom culture that you need to create is one where students have to think for themselves, be prepared to be stuck and know how to seek support, other than from the fountain of all knowledge – you!

Maslow's hierarchy of needs[3] reinforces this aspect of Mathematics teaching. Maslow's hierarchy of needs is a theory in psychology proposed by Abraham Maslow in his 1943 paper, 'A theory of human motivation' in *Psychological Review*. Maslow subsequently extended the idea to include his observations of humans' innate curiosity. His theories parallel many other theories of human developmental psychology, some of which focus on describing the stages of growth in humans. Maslow used the terms physiological, safety, belongingness and love, esteem, self-actualisation and self-transcendence needs to describe the pattern that human motivations generally move through.

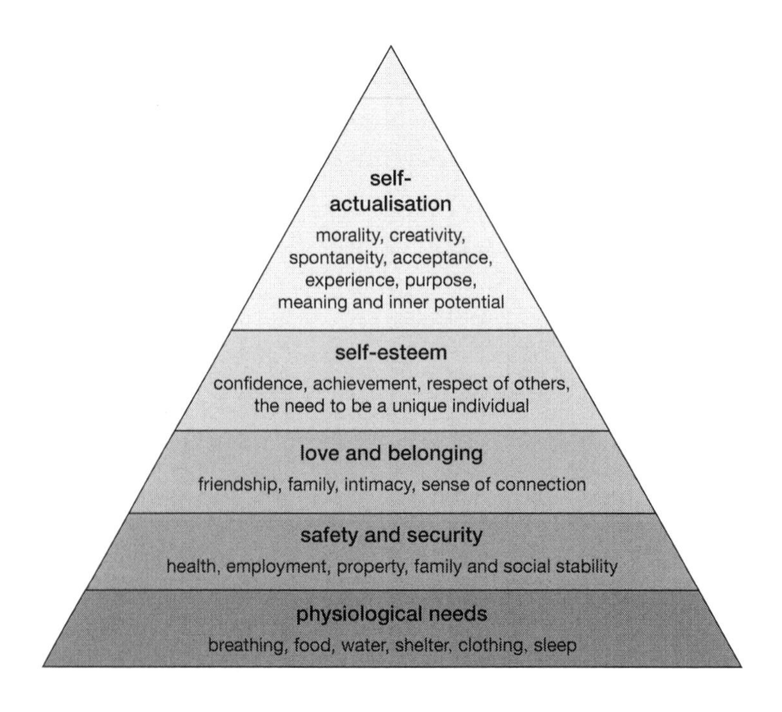

Maslow's hierarchy of needs is often portrayed in the shape of a pyramid, with the largest, most fundamental levels of need at the bottom and the need for self-actualisation at the top. Although the pyramid has become the de facto way to represent the hierarchy, Maslow himself never used a pyramid to describe these levels in any of his writings on the subject.

The most fundamental and basic four layers of the pyramid contain what Maslow called 'deficiency needs' or 'd-needs': esteem, friendship and love, security and physical needs. If these 'deficiency needs' are not met – with the exception of the most fundamental (physiological) need – there may not be a physical indication, but the individual will feel anxious and tense. Maslow's theory suggests that the most basic level of need must be met before the individual will strongly desire (or focus motivation upon) the secondary or higher-level needs. Maslow also coined the term to describe the

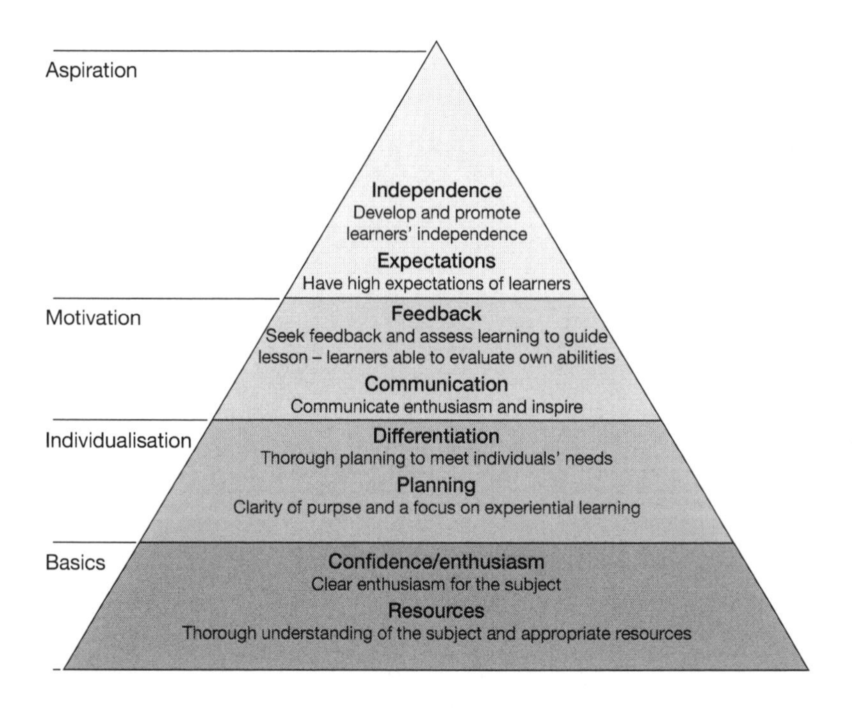

motivation of people who go beyond the scope of the basic needs and strive for constant betterment.

The pyramid, credited to Tony Davis, Ofsted inspector and learning and skills consultant, is not intended to be the chronology of how a lesson should progress. The Hierarchy of Teaching and Learning (HoT Learning) model was a by-product of the documentary film *ArtsWork*,[4] commissioned by the Adult Learning Inspectorate in 2006 to show the effect of outstanding teaching on learners' lives. It does not include such essentials as a clear introduction to the lesson, use of group activities, the final evaluation of learning etc. It is more to provoke a debate about the skills and attributes a teacher would need to acquire to move them from delivering merely satisfactory lessons to learning experiences that could not be missed.

So what would each of these components look like in the Mathematics classroom? From the bottom up:

Resources

Use of a range of resources is vital in the Mathematics classroom. The oft-quoted response to 'what did you learn in Mathematics today Jonny?' of 'page 132' is all too often the reality in some schools. In order to ensure coverage of the GCSE syllabus, some schools fall back on just one main resource for the delivery of the syllabus. A good Mathematics teacher will have a range of resources, from textbooks to worksheets, from YouTube clips to the use of interactive resources on the Internet, in order to utilise a range of stimuli for learning. Good use of resources will then require judicious selection of the most appropriate resource for the topic, style of task and group being taught.

With many schools opting for longer lessons, the range of resources used within each lesson is also increasingly important. In a 45-minute lesson, the use of one text and a range of questions will be sufficient to engage and instruct students. However, if teaching across a number of lessons or in one longer lesson, one single resource will not be adequate to sustain and motivate. Of course, it takes time to develop knowledge of a range of resources and build up that bank of expertise, but the only way to increase this when you start teaching is to keep looking at what is out there for each topic taught. There are many web-based resources for teachers these days, but all of them require thought about what purpose you are employing them for. Very rarely do I use something 'off the shelf'. I would always consider and adapt the resource to suit my own teaching style and groups.

Perhaps as a starting point, as planning in the first few years takes considerable time, take at least one topic a week where you thoroughly research what resources you could use. In that way, you will build up some ideas, without being overwhelmed for every lesson you teach. Remember to talk to more experienced colleagues as well: hearing what they use (and, indeed, what they have discarded) is invaluable training in those early years.

Confidence/enthusiasm

Whether or not you love your subject and feel secure in your own knowledge will come through in seconds to students. This should not scare you into thinking that you have to be the expert on everything, right from day one, but you must think about your delivery and whether or not it will instil confidence in those in your classroom. Planning is the key to confidence in your delivery and how you can explain to the students what they need to do.

Planning

We will deal with planning in more depth in Chapter 3, but suffice to say that thorough planning is at the heart of every good lesson. Students know when you are 'winging it' and when you have planned the lesson on the back of a postage stamp in the car on the drive to school. We all use times like these to generate ideas, but the best lessons come from detail of the progression of learning and development of resources that will support this.

Differentiation

Don't be fooled into thinking that, when teaching Mathematics, generally in ability groups, differentiation is unnecessary. Every class needs you to differentiate for different student needs. Consider the following few questions, all of which create their own differentiation challenge:

- Does your lesson have appropriate resources for those students who learn and understand conceptually, but struggle to write complete notes?
- Does your lesson allow for those students who can listen to your explanation and fully understand what to do, as against those who need to see the problem visually?
- Does your lesson consider completion of a range of questions, of increasing complexity, and will *all* students have to complete *all* questions?
- Does your lesson explicitly teach the skills of mathematical presentation, or does it make the assumption that all students will just do as you do, following a single worked example?
- Does your lesson allow for those students who were absent last week and missed a key explanation or concept?

Differentiation doesn't just mean planning for the more or less able, it means considering the needs of different students in your classroom and how they will learn best at all times. I have studiously avoided the mention of visual, auditory and kinaesthetic (VAK) learners, as I personally don't subscribe to planning your lesson with these learning styles in mind. I think it can be unhelpful to think about students in these three categories and to plan your teaching for their strongest style. Much research now suggests that students must work on their weakest element, not the strongest, if they are to make the most progress in the long term. What I *do* advocate is consideration of what will be the stumbling block in my lesson for different students, and how my planning seeks to minimise this.

Communication

This is strongly linked to the confidence/enthusiasm lower down in the pyramid, but discretely different. Communication is not just about what you say, but how you say it. Think about the following questions:

- What language will you use to explain challenging concepts?
- Will you use appropriate mathematical notation, or will you 'dumb down' language?
- How will you break down complex tasks into more manageable steps?
- How will you use diagrams to explain tasks and procedures?
- Will you explain the *why* as well as the *how*?

A good example of this could be how you might teach use of the quadratic formula for the first time. You will probably have already covered work on quadratics and their solutions, through factorising. You might also have done some work on the graphs of quadratic equations, and how these link to their solutions. So how might you teach the use of the formula? Consideration of what you teach and when is covered in more detail in Chapter 3. (This example, although making a point on communication, might also spark an interesting question in your head about the order of syllabus coverage and how topics build on one another.)

Assuming that I had done all of the above, I would start this topic with an equation that does not factorise and ask the class how we might find the solutions for such an equation. Does it mean that it does not have any solutions, because it does not factorise? How could we find out if this is the case?

I would live in hope that a student would suggest that we could draw the graph of the equation to see if it had any solutions.

Let's try $y = x^2 + 4x - 10$. The graph of this clearly does not factorise, and yet it does have two clear solutions. There must be a way of finding these.

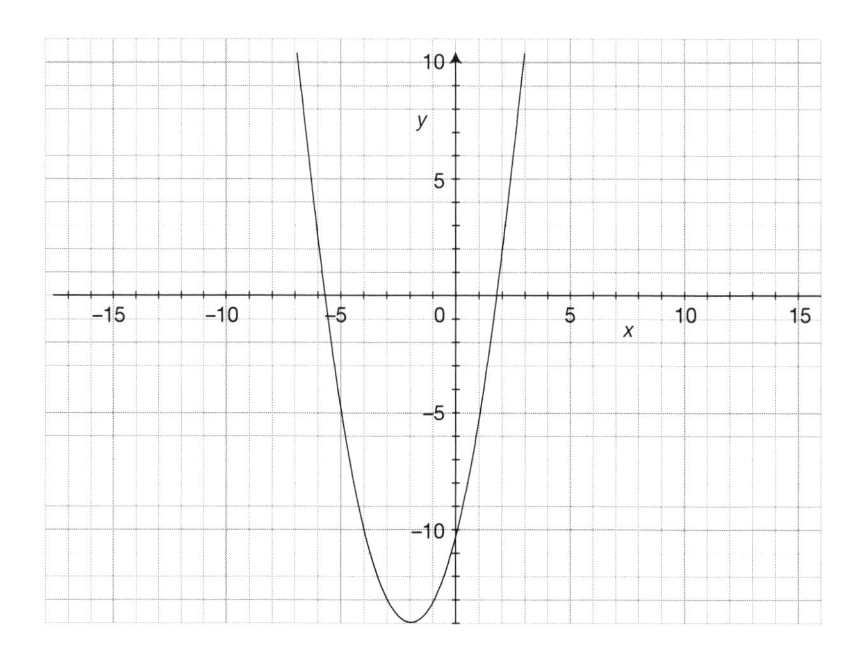

I would then go on to show the quadratic formula, using the form

$$ax^2 + bx + c = 0$$

I would find the solutions using the equation and then I would take the class back to our graph, to reinforce that the solutions that we had found matched the *x*-axis intercepts that we had learned when factorising. I have had many a debate with other teachers about whether or not, at this stage, you would show students the proof of the quadratic formula. I would always stress *yes*. Although some may struggle with the algebraic manipulation involved, this is the *why* and essential to a deeper understanding of the Mathematics involved.

Feedback

You will struggle to escape reading Dylan Wiliam's papers on assessment for learning. The paper that put Dylan Wiliam in the spotlight was 'Working inside the black box'.[5] Worthy of a read, as a booklet on this theme with a mathematical focus, is 'Mathematics inside the black box'.[6] The booklet suggests methods for teachers to interact more effectively with students on a day-to-day basis to promote their learning, for instance by using focused questioning techniques and careful analysis of pupils' responses.

The original paper, and many that have followed from Wiliam and many others, have unfortunately, in some instances, been interpreted into a myriad of lolly sticks and mini-whiteboards within the classroom. Although, if you walked into my classroom, you might well see the use of such devices to gain instant feedback from students, these alone do not make good assessment for learning. All such devices are worthless if they are not used to shape, adapt and appropriately change the direction of a lesson or sequence of lessons. What would be the purpose of students answering a question on a mini-whiteboard if their answers do not make me consider what I need to do next? The response to this will, of course, vary depending on the scale and extent of the misunderstanding, but respond I must.

Expectations

I am a firm believer that you must expect more of your students than they expect from themselves. This expectation must not be softened or muted for those who are less willing or less able. Take the simple example of homework. Would you expect less time and effort to be spent on homework by those in set 5 than by those in set 1? Disappointingly, often the answer is yes. This decrease in expectations just makes the lack of progress made by those students a self-fulfilling prophecy. Your expectations of students must be equal and challenging, irrespective of their background, response to the subject or desire to achieve. We are the professionals and we cannot forget that we are paid to teach young people and to help them to achieve. Sometimes, we can fall into helping only those who want to learn and forget that we are paid to help *all* to learn.

I remember distinctly bumping into a former student whom I had taught at GCSE. Aged 16 she lacked motivation, had a poor attendance record and did not engage well at school, and yet she was a bright young lady who was capable of achieving the very best GCSE results. Many others wrote her off, frustrated by her lack of commitment and drive. I taught her in a group of lively and vociferous Year 11 students who liked to discuss their work (and sometimes less relevant topics as well). I never expected any less of her than I did of anyone else, despite her lack of motivation. When I saw her again, some 10 years later, she came up to me to say, 'Thank you, thank you for never giving up on me and always expecting me to produce the goods'. I know that this isn't always easy, in the face of a challenging class or a group of students whose previous experience has been a negative one, but, as soon as you give up on them, they will, I guarantee, give up on you.

Independence

This is harder to create, the older the students that you inherit, but one critical aspect of any learning is requiring students to think. I know this sounds incredibly obvious, but there is a huge pressure on us all to get students to pass examinations at the end of the process. The result of this can be a spoon-fed diet of learning that ticks the boxes of the examination criteria. Ultimately, this approach will never be the most productive and never embed skills for life.

Independent learning, or the learning-to-learn journey, usually starts with active learning. Teachers create lessons in which students do a lot for themselves. Typically in groups, they solve problems, work things out, put pieces together, investigate, create, enquire and tackle stiff challenges, through activities where they have to think for themselves, talk about the problems that they face and, invariably, get stuck along the way. None of this is new; John Dewey was an articulate exponent in the early twentieth century, and many of the classroom strategies used are akin to Spencer Kagan's 'co-operative structures'. Through these active tasks, students practise social skills, organisational skills and all kinds of thinking, which stand them in good stead and help the learning process.

There are other benefits to active learning. If tasks are pitched in Lev Vygotsky's 'Zone of Proximal Development',[7] tantalisingly out of reach, students are mentally stretched. Chaklin[8] more recently expanded upon this in a work that took Vygotsky's theories further still. Students develop better thinking powers and also understand material more deeply because they have worked it out for themselves ('constructivism' – a hugely important educational truth – was one of Jean Piaget's key ideas 50 years ago and is currently promoted by Dr Chris Watkins, Professor John West-Burnham, John Abbott, Mike Hughes and many others). On top of the eventual acceleration in progress and deeper understanding, teachers and students alike report more enjoyment in lessons. So, as the pyramid suggests, this really is the icing on the cake.

Through the work of a teacher in my current school, we have created three learning posters and use these to explicitly reference the learning process and develop students' independent learning skills. Although the posters alone do not develop students' independent learning, alongside appropriately thought-provoking materials and lessons, they reinforce the need for this aspect of the best lessons.

Questions to think about
before learning

What do you want to learn today?

What skills do you have that could be useful in this lesson?

What are the signposts to your learning? (Must – Should – Could)

What do you know that might be useful?

When have you had to think like this before?

Questions to think about
during learning

What are you thinking about right now?

What connections have you made between what you are learning about now and previous learning?

What do you want to learn more about?

How have you got involved in the lesson?

What should you do to further your thinking?

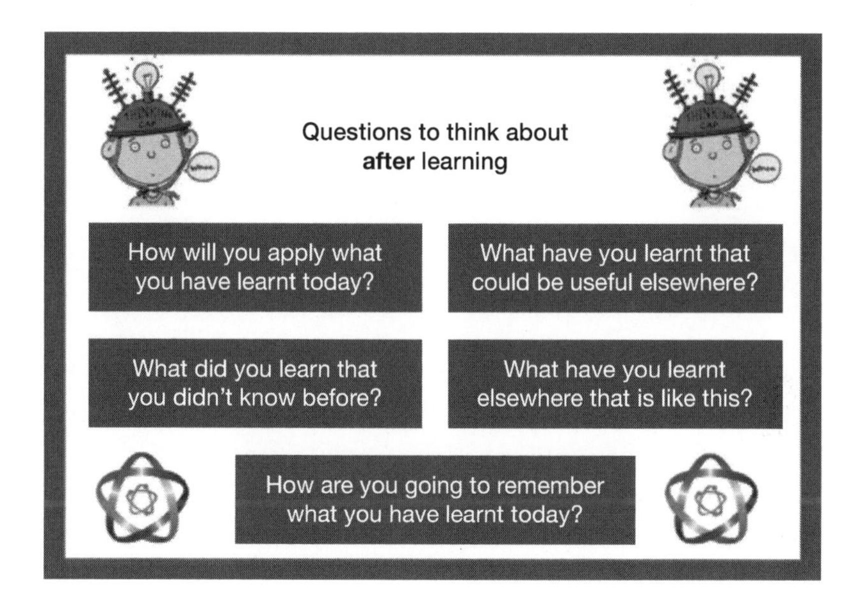

Tasks that encourage independence, reasoning and problem-solving skills

So, what types of task should you employ to develop this style of learning in your classroom? There are many investigational resources that you can find that give broader, more open-ended tasks that serve as a starting point for exploration of a range of mathematical concepts.

ACTIVITY: RICH QUESTIONS FOR INDEPENDENT LEARNING IN MATHEMATICS

We'll give one example. This example comes from NRICH,[9] a resource created at the University of Cambridge that aims to build rich tasks of exploration and discovery into an essential part of every Mathematics classroom. From this starting point, consider where the task might lead. What skills or topics could this lead into for a low-ability group or for a high-ability group? How might you use this as a starting point for a teaching concept? What would you expect to be the outcomes from time spent on this task by students? How would you know that they had made progress in the task?

How many different triangles can you make which consist of the centre point and two of the points on the edge?

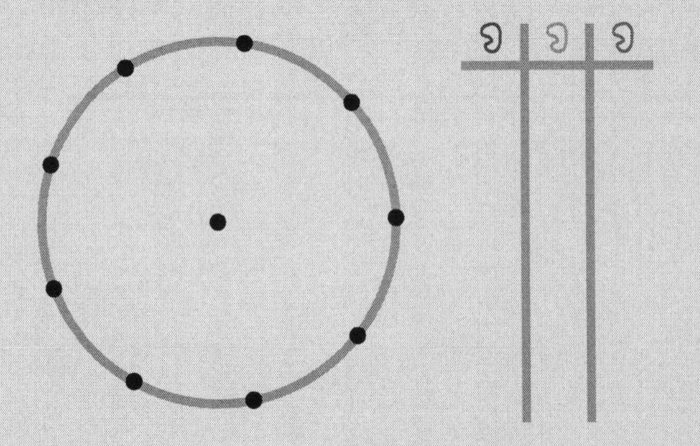

Further to the initial question in the figure:

- Can you find the angle at the centre point?
- Can you find the other two angles in the triangle?

Give yourself just a few minutes to think about where this starting point might lead. I shall give you some space in which to scribble some thoughts and I want you to consider your response at two levels, for a foundation-tier, lower-ability student and for a higher-tier, abler student.

Lower ability

Higher ability

See the Appendix for my responses.

I am sure that you will have found more possibilities. This broad starting point is a valuable tool for getting students to apply skills they have already learned and then develop and utilise these in a new area. They encourage mathematical discussion and develop skills of reasoning and independence.

For me, rich tasks should:

- be accessible to all as a starting point;
- be extendable and have a range of possible responses;
- be tasks that need students to make choices;
- promote discussion and communication;
- encourage 'What would happen if . . .' and 'Why can't . . .' questions.

Rich tasks can be longer and can delve into a range of mathematical teaching, or they might just spark the students to think about their use of a particular skill, like this rich task, which would make a simple starter:

> I've just been to the sweet shop. I spent 40p on sweets. I gave the shopkeeper 50p. I was given 10p change.
>
> What coins could I have been given?

This type of mathematical enquiry might be considered 'open Mathematics', whereas a list of twenty questions that practise a skill might be considered 'closed Mathematics'. There is, and I am sure there will remain, a debate as to whether a diet of rich or open tasks in isolation is the most appropriate mathematical diet. There is much to suggest that they build confidence and success in the classroom. My experience would suggest that they can also build failure and frustration in those who are challenged by Mathematics. As I will now go on to assert, I think there is as much a place for rigour and practice as there is for exploration and discovery.

What place is there for consolidation and rote learning?

I started this chapter implying that there was a debate to be had on this topic, and it seems as though, up until this point, I have advocated a classroom of group work, practical problems and discussion. This is the part where those in inspectorial positions who wish to watch one-off, exciting lessons where students beam with the light-bulb moment of discovery when they come to observe us, will have to leave the room. I shall say it loud and clear – *there is a time for consolidation and practice*. This is absolutely critical in your classroom. If they hadn't spent time training, would David Beckham have scored so many goals, Jonny Wilkinson converted so many tries or Steve Redgrave won so many gold medals? Of course not. So why do so many assume that, once a child has been taught or grasped a concept, it will stay in their mind forever, without practice?

Lessons from the Far East?

In Chapter 1, we looked briefly at the differences in performance between the UK and the Far East. One of the most noticeable differences – if you have the chance to observe this first hand, then take it – between Mathematics classrooms in the UK and the Far East is the expectation with respect to practice. In the Far East, time in class is mostly spent going through worked examples in front of the class. These are varied and numerous. They are worked through to give full solutions, showing the stages in methods and calculations. There is little discussion.

A journal written by Chap Sam Sim and published in *Mathematics Educational Research Journal* in 2007[10] looks at the features of a typical Shanghai secondary school lesson.

- Start of the lesson: the teacher revises a previous concept as a refresher to content learned either recently or some time ago.

This concept may be related to the topic being studied in the lesson today.

- Concept explanation: the new concept is introduced and, through a range of examples, it is demonstrated. These examples are varied and numerous and become increasingly challenging.
- Skill acquisition: students then spend a limited time within the lesson looking at examples that further fit the concept being delivered. Exercise questions have a significant variation in difficulty level. Students come to the board to demonstrate their working, with full methods. Other students listen attentively to the explanations of their peers and adjust their work accordingly.
- Engagement of students: students are called to the front to explain their work to the class. They rarely engage in discussion in groups while completing work. The atmosphere in the class is serious and orderly at all times.
- Plenary of the lesson: the teacher summarises the concept that has been covered and sets the work to be completed for homework. This homework will be further questions on the concept covered during the lesson.

This classroom and its atmosphere may not be dissimilar to many around the UK, but I think we would probably all agree that this seemingly didactic instruction was more familiar to our parents and classrooms of the 1960s in the UK than the modern, discussion-based learning and encouraged independence. The homework aspect is of great interest to me. This type of homework is set after *every* lesson and is expected to last at least half an hour on each occasion. Time and space are often available within school in the form of a 'prep' after-school session. There is an understood expectation that this work forms an essential part of the learning and will be marked by the teacher. Students complete the work without fail, and parents know that they have a responsibility to support their child in completing this learning. Education is seen as the key to your future, and a hard work ethic is expected in all families.

Interestingly, when you look at the allocation of teacher time in these Far East countries, it is very different from the UK. In the UK, a classroom teacher, with no other responsibilities, can expect to be teaching between 85 and 90 per cent of the school day. Preparation and marking take place outside this time and are expected parts of the job. In one Shanghai school, for example, the balance is very different, with teaching time in front of class closer to 30 per cent, and the rest of the working day and evenings expected to be used for planning, marking, intervention and support. Teachers generally only have two classes for which they are responsible. It is also worth noting that, at the primary stage, students have, on average, 45 minutes of Mathematics a day, and, at secondary school, this increases to around 1 hour. It might sound an easier life, but, with more than 4 hours of work set to classes with an average class size of thirty-five and expected to be marked and returned by the following day, it clearly is just a different pressure and allocation of time.

However, all this tells us that consolidation and practice must be at the core of their high standards and they are things that, I fear, many Mathematics classrooms have shied away from in recent years, in favour of problem-solving and discussion-based learning that are perceived to be more engaging. Don't get me wrong: as I have argued from Chapter 1, an enquiry- and discussion-based classroom will often increase student engagement, enjoyment and, hence, attainment, but this *must* be backed up by time for consolidation and practice. The ideal, in my mind, would be that this consolidation takes place at home, as in Shanghai. However, we must be realistic about the context in which we are working. The reality is that the expectations about completion of schoolwork at home in the UK are not the same as those of parents in Shanghai, and so we must allocate time for it within our classrooms. I would recommend that you, at least on a weekly basis, let your class complete questions, in near silence, to focus their attention on the critical practice element of this subject. It is this rigour in consolidation – call it *training* – that will embed skills and help students to retain new learning.

Interestingly, the emphasis in primaries in the Far East is also very prescriptive. The language used is very precise (I will explore this further in our focus on literacy in Chapter 6), logical reasoning and deductive thinking are a key focus, and the use of ICT is very evident, with mathematical software for geometry and graphical topics. Another interesting feature of primary Mathematics in the Far East is the continued use of some form of abacus, for a sound development of place value and number bonds. The example in the figure is commonplace in primary classrooms, and the seemingly unnerving ability of students to complete mental arithmetic later in their schooling is attributed to the use of this. Once again, it would appear that the rigour and practice of the use of this device at a primary level mean that students can recall what they did physically and almost reproduce the abacus visually in their heads when they calculate. This ancient tool of calculation seems to strengthen students' skills in fast and accurate Mathematics by enhancing and polishing brain functions such as concentration,

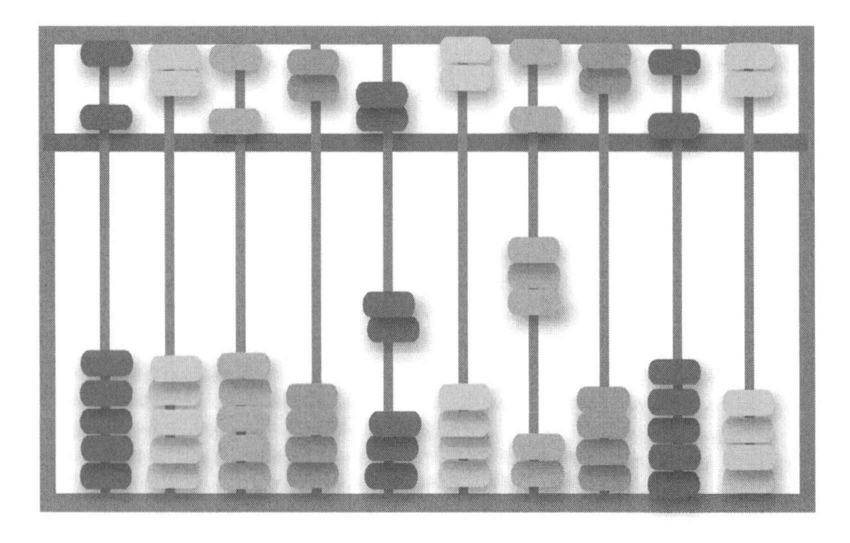

focus, listening, analysing, comprehending, imagination, retention, memory and perseverance. I don't need any more persuasion!

Some further key features in the secondary phase are continued precision of spoken language, a strict writing format, continued focus on logical reasoning and deductive thinking, a constant, fast pace to the learning, and little integration of concepts taught into daily life and contextual problems (this is more prevalent at the primary stage, but is not seen as necessary at a secondary level; Mathematics is enjoyed and used for the skills it develops).

Practice is an essential part of any learning, and especially Mathematics. However, practising problems can be tackled in different ways. Most textbooks include questions of increasing difficulty; however, they vary greatly in how quickly they become more complex. Some will provide dozens of questions of a very similar ilk; others include just a few 'settlers' before introducing more testing problems. Both have their place, depending on the topic and the nature and ability of the students. In my experience, students who are less confident in their ability feel far more at ease if they get some easier problems under their belt, before any variety is introduced. More confident mathematicians, on the other hand, can cope with the introduction of change more quickly. I'll come back to this variation and how important it is in your explanations and the work that you set in Chapter 7, when we talk about language.

So, what is the answer?

I hope that I have convinced you that the answer is that there is a place for exploration, discovery and open tasks, but these must be combined with rigour and practice. In essence, the best Mathematics classrooms have a *mixture* of the two. There will be a time for exploration and discovery, but there must also be time for consolidation and practice. A classroom that gets the balance wrong could make students less confident, less engaged, less able to reason for themselves, less able to retain what they are taught or less likely to learn in the long term.

TALKING POINTS

- How will you balance exploration and consolidation in your classroom?
- How can you build in structured opportunities for independent learning?
- Remember to think about the teaching and learning hierarchy of needs, and how activities in your classroom build confidence.
- Consider, when you start different topics, how you might allow an exploratory approach, using spatial reasoning and intuition for students to develop their own reasoning.
- How much time will you allow in the lesson for consolidation? Will this take place outside the classroom? If so, how much will you expect? And how will you mark and evaluate the work that students have completed?

Your thoughts

Notes

1 J. Boaler (2010) *The Elephant in the Classroom: Helping children learn and love maths.* London: Souvenir Press.
2 B. S. Bloom, M. D. Engelhart, E. J. Furst, W. H. Hill and D. R. Krathwohl (1956) *Taxonomy of Educational Objectives: The classification of educational goals. Handbook I: Cognitive domain.* New York: David McKay.
3 A. H. Maslow (1943) 'A theory of human motivation', *Psychological Review,* 50, 370–96.
4 See: www.artswork.org.uk/home
5 P. J. Black, C. Harrison, C. Lee and B. Marshall (2002) *Working Inside the Black Box: Assessment for learning in the classroom.* King's College London Department of Education and Professional Studies.
6 J. Hodgen and D. Wiliam (2005) *Mathematics Inside the Black Box.* London: NFER-Nelson.
7 L. S. Vygotsky (1978) *Mind in Society: The development of higher psychological processes.* Cambridge, MA: Harvard University Press.
8 S. Chaiklin (2003) 'The Zone of Proximal Development in Vygotsky's analysis of learning and instruction.' In Kozulin, A., Gindis, B., Ageyev, V. and Miller, S. (eds) *Vygotsky's Educational Theory and Practice in Cultural Context,* pp. 39–64. Cambridge, UK: Cambridge University.
9 See: http://nrich.maths.org/frontpage
10 C. S. Lim (2007) 'Characteristics of Mathematics teaching in Shanghai, China: Through the lens of a Malaysian', *Mathematics Education Research Journal,* 19, 1, 77–89.

3 For the love of Maths

I could not let this book pass without a chapter on why Mathematics is such a wonderful subject. You will undoubtedly encounter students and parents who will say, without any shame or hesitation, 'I hated Maths at school', 'Maths is boring' or 'What is the point in Maths?' I could have entered into a heated debate at this point about their narrow perception of the subject and how, in the case of parents, their negative opinion will be unhelpful to their child's motivation, how Mathematics has many core skills that, of course, are useful in life, but it also has an inherent beauty and teaches logic and problem-solving skills that are vital in any workplace. However, I have generally chosen to not argue with these people at the point the question is raised but have, as often is best, set out to prove by example that their perceptions are wrong, and that there are many reasons why Mathematics is fun, interesting and worthwhile.

I can guarantee that, for every student you come across who has an 'I can't' attitude to Mathematics, you will encounter four times as many students who get a buzz from getting a question right, solving a problem or, quite simply, completing a page of complex arithmetic or algebraic manipulation. Only at the end of last week, while running a revision class after school on Friday for some Year 10 students, did I hear a 'wow that is cool' after we had completed a complex changing-the-subject problem. Having started the session with fear and trepidation and questions such as, 'What are we going to do with it once we have rearranged it?' and 'What is the point in doing all that just to write it a different way?', at the end of the session they were buoyed and inspired that they could complete a multi-step problem with six or seven rearrangements and be sure that it was correct along the way. At the start of the session, I did not take time to answer their 'why' questions. I let the Mathematics thereafter do the talking. At the end of the session, I did have one trick up my sleeve, which was a number of equations that I had gleaned from their Physics teacher; to use them in their Science work, they had to be able to rearrange them confidently. I knew that, in a Science lesson in recent weeks, they had found this challenging, and it was limiting their learning. I had built my session around them enjoying the solution for its power and complexity in its own right, but we ended with a contextual example that they could see made the skill relevant and useful. This approach is how I endeavour to build a love of Mathematics.

We can't always and shouldn't have to always find a context to make the skill relevant. Students might question this, but, in fact, they do actually enjoy it as a subject in its own right and gain huge pleasure from seeing a page of completed problems. Mathematics can and does create a clear sense of achievement that can inspire and motivate students.

As Mathematics teachers, we must not be ashamed of loving our craft. There is a certain satisfaction when you come across a difficult problem and are able to conquer it, similar to a physical challenge. It is a subject that explains nature and the world around us. How cool is π, and its relationship with a circle? I defy you not to have had a sense of awe and wonder when you discovered the relationship $ei^{\pi} = -1$. How can two irrational numbers and an imaginary number combine to create something so complete? These miracles of Mathematics pop up in all areas of life, from radiation heat exchange, to compound interest, to pine cones in the wood. Mathematics can give you a sense of there being a higher power in the world and a divine order.

Of course, while teaching a lower-ability Year 8 group, you are unlikely to discuss these advanced areas of mathematical wonder, but you can still create a love of number, space, structure and order. I have never met a student who doesn't actually enjoy the order and methodical nature of Mathematics. Even creative types can gain pleasure from calculations and problems and have a sense of achievement from their work.

So, how do you foster this love in your students? First and foremost, by displaying your love of Mathematics and excitement at different topics. Don't be bashful about saying 'I love how this works', 'Doesn't that page of algebra look cool' or 'Isn't that formula clever?' Students need to see that you love the subject and that, in turn, they will too. Second, use puzzles and games to make students think about problem-solving that isn't in the curriculum. Most students love puzzles. There is a wealth of problems that you can use as starters or even as a 'problem for the week' on your classroom wall, to create an interest in things that are taxing. When I used to teach in my own classroom, I would have a problem on the wall each week. It wasn't part of the curriculum and could be answered by any student in any year group. It was amazing how much discussion the problem generated across all my classes, and students would often return to the room on Friday lunchtime, to see the solution before they left for the weekend.

ACTIVITY: THE WONDERFUL WORLD OF PROBLEM-SOLVING

This is one such problem that I am sure you will have come across.

Cannibals ambush a safari in the jungle and capture three men. The cannibals give the men a single chance to escape uneaten.

The captives are lined up in order of height and are tied to stakes. The man in the rear can see the backs of his two friends, the man in the middle can see the back of the man in front, and the man in front cannot see anyone. The cannibals show the men five hats. Three of the hats are black and two of the hats are white.

Blindfolds are then placed over each man's eyes, and a hat is placed on each man's head. The two hats left over are hidden. The blindfolds are then removed, and it is said to the men that, if one of them can guess what colour hat he is wearing, they can all leave unharmed.

The man in the rear, who can see both of his friends' hats but not his own, says, 'I don't know'. The middle man, who can see the hat of the man in front, but not his own, says, 'I don't know'. The front man, who cannot see *anybody*'s hat, says, 'I know!'

How did he know the colour of his hat, and what colour was it?

In Chapter 8, I will look at how you can use broad questioning techniques in your classroom to build confidence. Giving students opportunities to feel better about their work is at the heart of creating a love of Mathematics. Of course, a student who struggles and feels that they cannot achieve will not enjoy being in their Mathematics lessons. So, at the heart of students learning to love Mathematics has to be their having the confidence to try different things. In Chapter 4, on planning, we will cover the need to plan the curriculum so that it has coherence. Taking time to come back to concepts and constantly 'review' previous learning will build confidence and students' ability to recall. I will also go on to talk about how to plan questions with variation, the idea that, when you teach a skill, you consider carefully how your questions build knowledge and confidence.

All of these aspects are integral to building confidence in students and, hence, giving them a love of the subject.

TALKING POINTS

- How will you bring your love of Mathematics into the classroom?
- What activities could you do with all classes that promote a love of problem-solving?
- How will you build confidence in your classroom, so that students gain a sense of achievement and, hence, a love of Mathematics?

Your thoughts

4 Planning essentials

This chapter will look at short-term and long-term planning, considering how each topic taught builds to a coherent Mathematics understanding, with reasoning skills that cross different content. We will look at how you plan across the whole year in medium-term blocks, using knowledge of the Mathematics curriculum and with an eye on measuring progress, and then the nitty-gritty of planning a single lesson.

As we have already seen, the breadth of the Mathematics curriculum is vast, and it covers a range of topics and skills that can confuse and baffle students.

Planning your curriculum and considering what you will teach when and how are essential. You need to consider how different topics interlink, and what prerequisite knowledge each topic area will require. By starting to understand this complex web, you will explicitly help students to make those links themselves and, hence, build strong foundations of their understanding.

I have always worked with a three-tier plan.

Tier 1: long-term planning

Tier 1 should give an overview of the year and a glimpse of the topics to be covered.

Planning essentials

Each school will (should) create an overarching plan in order to organise and structure the year. Why is this important?

I would expect all schools to have at least this level of framework that all the teachers are working with. This is not about creating robotic classrooms where all teachers are doing the same thing, but, by teaching the same topic at the same time, you are more likely to discuss strategies and share resources. If you have ability-set groups (most likely, though not perhaps in Year 7), it also means that students can move groups without having missed chunks of work taught in one group but not the other. These reasons are all logistical and compelling, but, at the heart of a clear plan, there should be a rational argument for the benefit of Mathematics learning.

So, more importantly and more relevantly, as someone starting out in their teaching career, why is it important mathematically to have a clear plan for your teaching?

This Year 7 example is taken from the MEP[1] scheme of work. This is an overview of the academic year. This is just one example that you might see within schools.

Look at the range of topics in the plan. The scheme covers all areas, from number and algebra to shape and data handling. This could seem an unconnected array of topics to students, and yet they are subtly linked. Consider how the foundation skills of arithmetic, with decimals and fractions, will come into play when teaching probability later in the year. The earlier topic of areas of rectilinear shapes links to volumes of cuboids and triangular prisms, again at a later stage. Graphs and directed numbers on a number line at first glance seem a stand-alone topic in this Year 7 scheme, and yet, combined with the skills of linear equations and nth terms, will clearly build a solid base from which to be able to tackle linear (and eventually more complex quadratic and other) graphs and equations. It is no coincidence that this plan teaches the skills of basic arithmetic before entering into area and perimeter. Of course a child will not be able to tackle the concept of perimeter as the addition of the lengths of the shape, if they cannot do basic arithmetic.

Week	Topic
1	Logic – two-way tables and Venn diagrams
2	Arithmetic: place value – whole numbers, decimals
3	Graphs – directed numbers on a number line, coordinates
4	
5	Arithmetic: addition and subtraction of decimals – money
6	
7	Angles – basic geometry
	Half-term
8	Arithmetic: multiplication of decimals – whole numbers and decimals, money
9	Number patterns and sequence – nth term formulae
10	
11	Arithmetic: division of decimals – whole numbers and decimals, money
12	Area and perimeter: squares, rectangles and triangles
13	
14	Revision of term 1 topics: consolidation
	Christmas
15	Arithmetic: fractions – fundamental concepts
16	
17	Data collection and presentation – collection, organisation and display
18	
19	Arithmetic: revision – recap the four rules for whole numbers, decimals and money
	Half-term
20	Searching for patterns – nth term for geometric sequences
21	Time, timetables and mileage charts
22	Arithmetic: negative number – the four rules
23	
24	Algebra: linear equations – solving and creating
25	
	Easter
26	Arithmetic: decimals and fractions – conversion
27	Discrete quantitative data – organisation and analysis
28	
29	Scale drawing – lengths and angles
30	Arithmetic: fractions – the four rules
31	
	Half-term
32	Probability of one event – including addition law
33	
34	Volumes – cubes, cuboids, triangular prisms, capacity, density
35	
36	Review of the year – consolidation
37	

The other interesting aspect of this plan is how time is given for consolidation and recapping. This is essential in any planning in Mathematics. The range of topics covered can often seem disconnected, and the pace of coverage means that they can pass in a blur. It is essential that time is given, either briefly as starters within lessons, or more lengthy periods for consolidation and recapping.

So, when you consider your planning, there are two key things that you need to think about:

- What do the students need to know before they can tackle this?
- Where will this topic go at a later stage in the curriculum?

This takes us neatly to Tier 2 planning.

Tier 2: medium-term planning

Tier 2 involves consideration of what each topic entails and the objectives that you wish students to reach in the expected time frame.

Don't get too bogged down with this. I don't mean that, when preparing to teach linear equations with Year 7 for the first time, you consider how they might tackle the mathematical modelling in linear programming, *but* you must consider the skills that you are developing that will be used further along the line.

There has been much work at primary school level to address this very aspect of teaching Mathematics over the last few years. Tricks and methods that work in simple scenarios, when scaled up to secondary level and more complex skills, just don't hold true. Here are just three examples from primary level:

> When you times by 10, just add a 0.
>
> 5 − 7 = ? You can't take 7 from 5, or you can't take a big number from a small number.
>
> 6 × 8 is 6 lots of 8.

The last example is an interesting one, as you start to think, why can you not say that 6 × 8 is 6 lots of 8? The flaw here is a subtle one. Although it does the job, teaching the children that '×' means 'lots of' is not entirely accurate, because this terminology suggests that the calculation needs to be carried out in a certain order, and multiplication is commutative. A preferred statement would be 'it is 6, 8 times', to emphasise the repeated addition.

Let's take an example from our Year 7 scheme – Algebra: linear equations. Many students have a very good, intuitive understanding of basic linear equations. Most teaching starts with consideration of something more concrete than letters as variables.

For example:

So how much is one can of beans? Most students will be able to say 5 (using either addition of 5s or multiplication). Some may have understood the law of inverse operations at play and will have calculated 15 ÷ 3, but most will have considered what multiplies by 3 to make 15. I hear you cry that the latter is not an error, and that this child has understood perfectly how the pseudo equation works. The problem lies when this 'equation' becomes more complex.

How does the child who has used the 3 × 5 concept solve this equation?

$$3x = 13$$

This has the complexity of an understanding of algebraic expressions and x representing the can of beans, but, this aside, this problem is, of course, more complex, because the solution is not a whole number. Only the child who understood the inverse operation of divide will successfully solve this problem quickly. This, of course, then builds when students have to rearrange formulae and solve algebraic fractions and more complex equations. 'I can see it because I know that $3 \times 5 = 15$' just doesn't help.

I hope that, from this brief example, you can see why planning is so important and how you must consider what you are teaching, in the topic, but also for the future use of such techniques.

You will build up the knowledge of how the topics link together in time, but it is worth consideration when planning your lessons in those first few years. William Emeny[2] shared with us a visual representation of the Mathematics curriculum and how it all links together.

The figure shows how complex the many links are, and why this is essential consideration when planning.

William Emeny[3] explains how these links could form a basis for the bedrock of skills that students need to master at a younger age. In rank order, the most important topics for students to master, based on the number of topics for which they are prior knowledge, are as follows:

- multiply and divide whole numbers (this is prior knowledge for 90 topics);
- add and subtract whole numbers (73);
- BIDMAS (50);
- multiply and divide decimal numbers (43);
- understand place value and identify the value of digits in a number (38);
- add and subtract decimal numbers (34);
- multiply and divide negative numbers (34);
- write a fraction in its simplest form (29);
- round to decimal places (29);
- substitute into an expression or formula (26);

Multiply and divide whole numbers
Add and subtract whole numbers
BIDMAS
Multiply and divide decimal numbers
Add and subtract decimal numbers
Understand place value and identify the value of digits in a number
Round to decimal places
Round to significant figures
Multiply and divide negative numbers
Add and subtract negative numbers
Put a number on and read a number off a number line
Plot and identify coordinates
Extract data from lists and tables
Equivalent fractions
Convert between fractions, decimals and percentages
Fraction of an amount
Percentage of an amount
Multiply and divide by powers of 10
Two letter notation for a line and three letter angle notation
Substitute into an expression or formula
Expand and simplify single and double brackets
Collecting like terms
HCF and LCM
By William Emeny
www.greatmathsteachingideas.com

- add and subtract negative numbers (24);
- put a number on and read a number off a number line (23);
- round to significant figures (23);
- two-letter notation for a line and three-letter angle notation (21);
- use a calculator to evaluate complex calculations (20);
- plot and identify coordinates (19);
- use and calculate with index notation, including squares, cubes and powers of 10 (19);
- equivalent fractions (18);
- express one number as a fraction of another (18);
- extract data from lists and tables (16).

Sixteen of the top twenty topics are number and could be summarised as four operations, BIDMAS, place value, rounding, negative numbers and basic fractions. This will come as no surprise to the experienced Mathematics teachers out there.

Other subjects, such as history, take different topics or time periods but build the same skills. Some say that Mathematics is different and that the skills vary from one part of the curriculum to another. I would wholeheartedly disagree with that assertion and think that, through the different topics in Mathematics, we are constantly building some core skills. It is vital that, as you move from one topic to another, you keep coming back to those core basics: working with operations, dealing with negatives, algebraic manipulation, implications of BIDMAS etc. This is as well as reminders about things learned weeks or months previously. You need to think about how you will weave in previously learned topics – starter activities that use a skill learned last term or last year, a reminder of concepts or skills that are prerequisite as you start a new topic, or questions as part of homework that require knowledge learned some time ago. This constant review of skills will help to embed them in the students' minds, will make the students use their own resources (exercise books, textbooks, revision guides) to look up how they apply the skill before they ask you for the answer and, hopefully, will make those neural pathways stronger through regular recall and use.

Tier 3: short-term planning

The detail required for a single lesson

The third tier is your detailed lesson planning for each topic area, with a range of activities, timescales and forms of assessment. Different training colleges will ask for different forms to be completed to show your planning. What form this takes will vary from school to school and institution to institution. Here are two very different examples of the level of planning that you could employ at the Tier 3 stage. Of course, this Tier 3 planning will relate to Tiers 1 and 2, as it will have necessitated a thought process about what comes before and what comes after each of the topics being covered.

The 5-minute lesson plan

The 5-minute lesson plan, first developed by Teacher Toolkit,[4] has been much championed as a way of considering all the elements needed in an outstanding lesson. I think, for the experienced practitioner, it is a good visual reminder of different aspects of the lesson being planned. However, I think that it lacks the structure necessary when you are starting out in your teaching career.

Hence, the following is a lesson planning sheet, or something similar, which I would advocate at the start of your teaching career (or even later, when you have formal observations and want to share with the observers your planning thought process; remember, the most important aspect of your plan is that it shows that you have spent time thinking about the lesson structure and content, not about the paper that it is eventually written on).

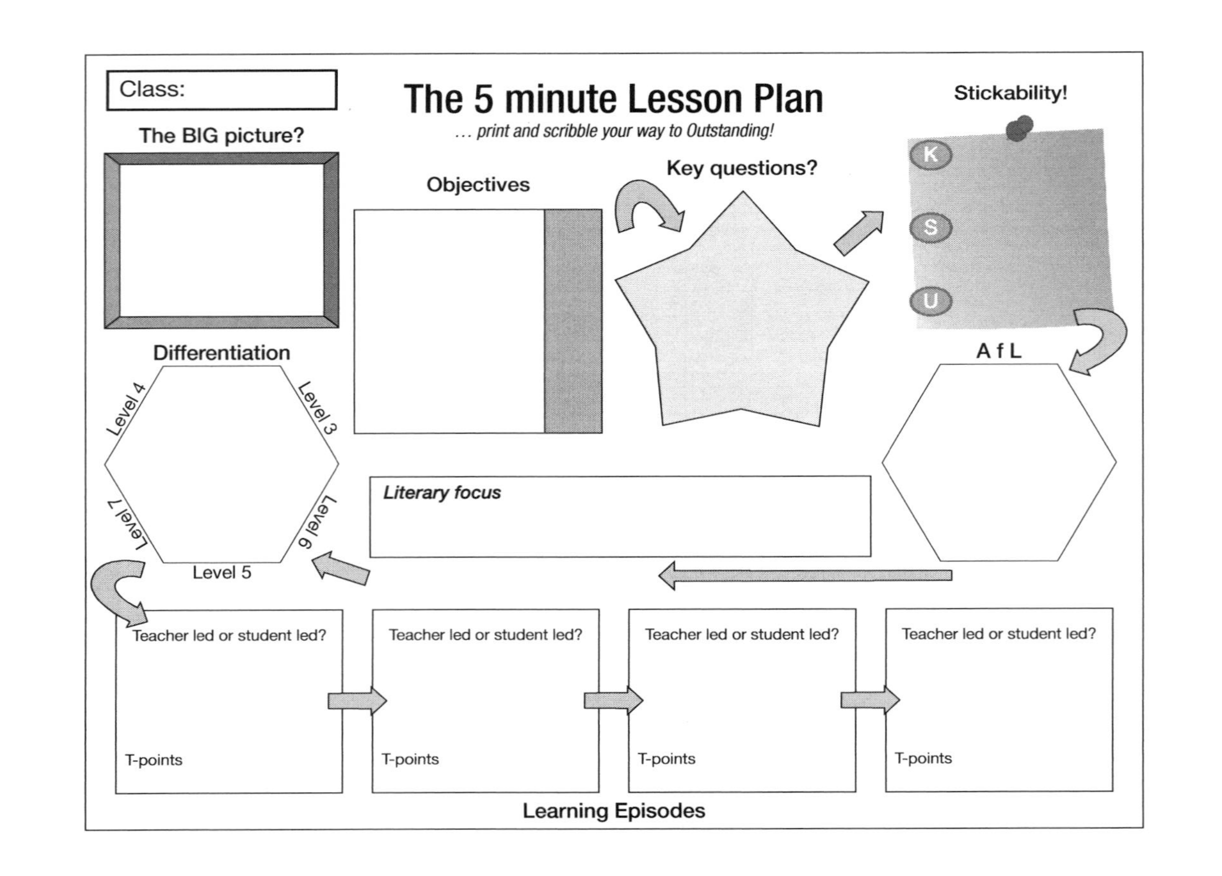
Class:
The 5 minute Lesson Plan
... print and scribble your way to Outstanding!
Stickability!
The BIG picture?
Objectives
Key questions?
K
S
U
Differentiation
Level 4
Level 3
Level 7
Level 6
Level 5
A f L
Literary focus
Teacher led or student led?
T-points
Teacher led or student led?
T-points
Teacher led or student led?
T-points
Teacher led or student led?
T-points
Learning Episodes

Lesson plan

Subject: Mathematics	Teacher: J Upton	Date:	Lesson
Class:	No. of students: M F		Setted 3 out of 6

Learning objectives (students will learn to do):	**Context** (scheme of work/links to sequence of lessons):
Recognise and draw graphical regions for inequalities. *We will be using a graphical tool to speed up the number of graphs that the students see in order to spend time thinking about the implications and conclusions rather than just drawing.* *ALL – draw and recognise an inequality for a simple region e.g. $x > 2$* *MOST – draw and recognise simple non-horizontal regions* *SOME – start to combine regions with multiple inequalities*	*Students have discussed what inequality symbols mean and represented these on a number line. They have solved inequalities – using previously learnt techniques for solving equations. I have encouraged students to think about what the inequality means, not just applying a process.* *This is the starting point of what a graphical inequality looks like and the intention is that next lesson students will move to drawing complex three inequality regions.*

Student groups (who is … for ability relative to the group rather than the whole year)			
SEN	**Lower ability**	**Middle ability**	**Upper ability**

Timings	Learning and teaching activities	Resources
5 mins	*Starter – reinforcing and recapping mathematical language associated with topics learnt this term*	*Sheet*
1 min	*Recap what we have learnt so far about inequalities and the next step for today*	*Powerpoint*
5 mins	*Give students an introduction to the task for the lesson to include:* *How to use OMNIGRAPH (students have used once before)* *What the graph will show for an inequality* *How to draw a 'sketch'* *What is the task?* *What are the aims of the task?* *Tell students the pair they are going to work in (all complete a sheet but are expected to collaborate with their partner for support – pairs are ability chosen to push the most able and encourage their high shared dialogue)*	*Sheet* *Omnigraph* *Computers*
30 mins	*Pupils then move to the computer room (leave bags in room) take pen/highlighter and sheet to make notes on (given out)* *There are four stages of complexity of the graphs that students are asked to draw their graphs and write their conclusions on the sheet*	
10 mins	*Return to the classroom to share some conclusions and 'test' their theories with the mini whiteboards*	*Whiteboards* *Pens*

Homework

Sheet that asks students to draw three inequalities (of increasing difficulty) and then to try the challenge to draw the combined area that fits all three regions

Progress measure for each ability group:	**Support staff role:**
Upper – start to combine regions with multiple inequalities *Middle – draw and recognise simple non-horizontal regions* *Lower – draw and recognise an inequality for a simple region e.g. $x > 2$*	

Planning essentials

There are five aspects of this plan that I would want you to think about.

Learning objectives

Make sure that these focus on what the students will learn, not just what the students will do. The latter could be very task-based and does not help to think about the learning process.

Context

This comes back to our Tiers 1 and 2 planning discussion and making sure that, even for an individual lesson, you have thought about the before and the after of the learning involved.

Student groups

These are useful, alongside a seating plan, if you are being observed, so that the person watching, if unfamiliar with the class, can identify key cohorts within the classroom. However, it is there in bold at the top of the lesson planning sheet as a powerful reminder of who you are planning this learning for. What differentiation will you need to consider for each group, or, indeed, for individual students in each group?

Teaching and learning activities

What preparations have you made and what will you use and when, throughout the lesson. I would advise a timeline that sits alongside this, that forces you to consider how long each activity will take that you have set. In that way, you increase the pace within your classroom and ensure that you think about time needed for any transitions between tasks. The only health warning that I would stress with such a timeline is that you don't stick to it so rigidly that it does not account for how the class approach the task and how well they do.

Progress measure

This should strongly remind you that, if the lesson is to have worth, then you will know at the end of it how well the students have done and if they are ready for you to move on. Be careful that such progress measures are not tokenistic and gimmicky, but that they demonstrate deep learning. It may be that the measure of progress will come at the end of a sequence of lessons, but be confident in your mind that you can answer the questions:

- Did they learn what I expected them to today?
- How do I know?

Appointments and that first job will be talked about later in Chapter 10, but planning and the support given at a school for this will be essential to making that first year more bearable. When you visit a school prior to or at interview, does its planning guidance include Tier 1, Tier 2 or Tier 3?

TALKING POINTS

- Consider those three tiers of planning: short, medium and long term. What does the school you are working with have in place already?
- Plan using the three tiers: long-term first, topic outline next, and, finally, detailed lesson planning. The temptation is to start at the detail and then just find your way, lesson by lesson.
- Consider how much time each topic will take and how much your classes will achieve in a lesson, week, month or term. You'll get this wrong to start with, but gradually your 'guesstimate' will become more reliable.
- When will you take time to stop and look at the overview, before diving into the detail of lesson-by-

lesson planning, so that your lessons have coherence and flow across the year(s)?

- How will you build in constant review of previously learned skills?

Your thoughts

Notes

1 See: www.cimt.plymouth.ac.uk/projects/mep/
2 See: www.greatmathsteachingideas.com/2014/01/05/youve-never-seen-the-gcse-maths-curriculum-like-this-before/
3 See: www.greatmathsteachingideas.com/2014/01/05/youve-never-seen-the-gcse-maths-curriculum-like-this-before/
4 See: http://teachertoolkit.me/the-5-minute-lesson-plan/

5 Assessment essentials

This chapter will consider what is and isn't good assessment in Mathematics. It will consider how often and how detailed this feedback should be, and different ways of assessing to give you different information that will support learning. It will consider the dirty word of *marking* and how this sits within a good assessment routine. It will tackle feedback as an intrinsic part of assessment and consider possible routines that will give a structure to creating a learning dialogue within every classroom.

Why assessment matters

Two basic assumptions of mainstream thinking on assessment are that individuals possess attributes (e.g. knowledge, understanding, skill, ability) that are discoverable and measurable, and that the primary purpose of assessment is to discover and measure these attributes. These assumptions are, perhaps, even stronger in Mathematics than in other subject areas: there are only right or wrong answers; you either know the right answer or you don't. Although the final GCSE examination, A level module or whatever other external hurdle is in place at the time is undoubtedly going to continue to fit this model of assessment for many years beyond my lifetime, there is no reason why all assessment that takes place within my classroom needs to fit this model as well. In fact, I would assert that learning and assessment for learning will be much the poorer if it does.

If you have ever read an Ofsted report from a relatively successful school (let's say at least 'Good' in the rankings), then you are highly likely to have seen a statement similar to the following: 'Improve the quality, consistency and regularity of marking and feedback across all subjects.'

Let me state for the record now that I *hate marking*. I find it the most repetitive and boring part of the job of teaching, *but*, and it is an enormous 'but', if you don't do it well and don't keep on top of it, you may well lose your sanity and the respect of your students and can wave goodbye to those target grades for your class.

Gosh, that all sounds a bit drastic for a bit of marking, doesn't it? Surely, I hear you cry, the most important aspect of teaching is the quality of the lessons that I deliver. The need for my lessons to include pace, challenge, variety, awe and wonder, depth of subject knowledge, support and attention to individual learning needs, all of which will help my students to make progress. *Exactly*. You are not wrong: your lessons must aim to deliver all these things and more, but how will you do that if you never mark any work? Never provide personalised feedback? Never look to see if your students have understood in depth the awe and wonder that you have been imparting to them?

Of course, I am being rather provocative, but I think that *assessment* has to be at the heart of good lesson *planning*.

The key aspects of assessment

So, what are the key aspects of assessment? I use PRAM for consideration of assessment:

- personalised
- regular
- active
- meaningful.

I shall tackle these in order and demonstrate how, if you stick to PRAM as a guide, your students will make more progress and engage more in learning.

Personalised

Students need to feel as though their work has been marked, and feedback has been given that applies to them. There will be many occasions where you do whole-class feedback from an assessment, but, as much as possible, keep it personal. How can you do this in a class of thirty? It is of the utmost importance that, wherever possible, you make feedback personalised and involve students in this process. I shall give some concrete examples later when talking about feedback, but for now, just consider how you will get beyond the 'one size fits all' feedback, where you go through some common mistakes at the start of the lesson in a blanket, everyone-must-listen-to-everything, fashion.

Regular

This is perhaps the easiest to explain but the hardest to achieve.

Routines are so important. If you have read *The Power of Habit* by Charles Duhigg,[1] you will know how much of an impact routines and regularity can have on success. Duhigg argues that, 'Habits are at the root of how we behave. They can be changed, if we understand how they work.' Once you break a habit into its constituent parts, he says, it makes it easier to control. He breaks it into three: the cue (or trigger), routine and reward.

For concentration during explanations in class, the trigger is teacher talk, the routine is distraction, and the reward is the chance to drift off, daydream or distract others.

It seems harder, because, although you keep the cue of explanations, you need to change both the routine (from distraction to concentration, or not *SLANTing* to *SLANTing* (SLANT stands for '*s*it up straight, *l*isten hard, *a*sk and answer questions, *n*od sensibly and *t*rack the speaker' – track, as in, turn towards whoever is speaking and look at them)) *and* the reward (from drifting and daydreaming to learning). This might also require improving the quality of explanations in the first place; in other words, changing cue, routine *and* reward; not just breaking a habit, but making a completely new one. There is a range of these 'routine changes' that could be applied to the classroom:

Negative habit	***Positive habit***
Slouching	Listening
Getting distracted	Concentrating
Shouting out	Taking turns
Giving up	Persevering
Forgetting equipment	Bringing equipment
Forgetting homework	Completing homework
Making excuses	Avoiding excuses
Feeling demotivated	Becoming self-motivated
Avoiding responsibility	Taking responsibility
Blaming others	Avoiding blame
Littering	Tidying up
Being rude	Being polite
Being mean	Being kind

When we consider planning and ensuring good behaviour, you will see how many of these 'routines' come into play, but let's focus on assessment and regularity with Duhigg's mantra in mind.

Students like to know how they are doing and they appreciate feedback, good and bad, if they feel supported in addressing any weaknesses. If this happens on a regular basis, it forms part of the routine of their learning, and, after a while, they look for that feedback as a critical part of their progression. What does this mean in practical terms? Have key assessments every few weeks (most organised Mathematics departments have shared assessments at least every half-term, where the results are then stored to allow examination of performance across groups). If these are not in place in your school, then, as a newly qualified teacher, you will find it hard to find the time to write these for a range of classes, but an assessment that brings together four to six weeks of work is essential for students' progression in the long term.

Alongside these longer assessments, probably conducted in lesson time, there must be regular assessment of classwork and homework. Build in a structure to your week. Let's say you teach Year 7 for four lessons a week.

- Monday, lesson 1
- Tuesday, lesson 2
- Wednesday, lesson 3
- Friday, lesson 4.

Set your homework on the same day every week and set it every week. That sounds so obvious, but both you and your students will benefit from a routine. I would always leave at least a day between setting and collection. So, set on a Tuesday and collect on a Friday; that way, students can see you if they have any problems and can ask questions if they wish. Make your day for collection of homework a day when you have some time after the lesson for any punishment. If it is not handed in, keep the student in there and then to get it done. If you set the tone with this from the start, you will soon find they get the work done on time. You might find that there is such a routine and expectation within your school or department. If so, great; if not, set your own. We all have other things in our lives, students and teachers alike. If a student knows your routine, they can plan their activities around it. You are far more likely to get work completed this way, than if you set it once in a blue moon and on a different day every time.

However, with this regularity comes the marking. Take the work in, get it marked and get it back to the students as soon as possible. You can lose face every now and then with, 'I'm sorry I haven't marked it yet', but, if you want to set the standard for them, you have to set it for yourself as well. If I collected homework on a Friday, it would be marked and given back to them on a Monday.

This also creates a routine to your lesson structure – Monday's return of homework will always start with a review of skills that students need to practise. Your assessment will have informed your planning, and you will have in mind what things you need to come back to (with individuals or the whole class) as you move through the topic, or return to it in the future.

Active

All assessment has to feel as though it involves the student. It is not just the means by which we put marks in a book to keep others happy and inform parents. Just as we discussed when considering the teaching and leaning pyramid of needs, assessment has to encourage independent learning skills and the involvement of the learner. Varying the types of assessment task will be a good starting point for this, as will ensuring that not all assessment comes at the end of a topic. Why not assess part way through, give students feedback, and then ask them to address their weaknesses in the second half of the unit of work?

Meaningful

What should an assessment contain, and how do we make it *meaningful*? I would always advocate that assessments should cover a range of skills, so that the students are not being assessed on one skill in isolation. This, of course, will become more complex as you go higher up the years, but it is essential that these formal assessments build towards modelling the need to move from one topic to another in quick succession. If your planning structure is a sound one for building skills over a range of different topic areas, students will also start to see the links between topics as you assess across them.

Types of assessment

Most often in Mathematics, you will see a style of test as a form of assessment, but I am going to suggest that a range of assessment methods is helpful to build confidence and skills in memory and recall, and to know how you need to help students prepare for that final pinnacle examination.

Formal tests

These generally include a range of questions across a number of topics, in either a module of work or across a whole term or year of work. There are many examples of software that will help you to write these types of assessment, drawing from old A level, GCSE or Key Stage 3 banks of questions. These have their place and are absolutely necessary to allow students to experience the type of assessment that they will face at the end of the course. However, they don't always tell you everything you need to know about a student's understanding, and they perhaps rather tell you who can remember the best and cope under pressure.

Consider these other forms of assessment that can be used to build skills to be successful in those more formal assessments.

Open-book tests

Students look at you rather oddly, the first time that you suggest this. You set a series of questions that, perhaps, are very similar in style to those in our formal testing. However, the one key difference is that students are allowed access to their notes when they sit the test. This might be their classwork book, a revision guide or pre-prepared notes (I'll come to this in more detail in a minute). You, of course, need to consider the length of time that students have for the test and the time that referencing material might take them. This extra time is significantly outweighed by the lesson that this teaches the students – that their notes do matter! The quality of what they record on a day-to-day basis will be dramatically enhanced if they then are able to use these in an assessment setting. Full working and explanations of techniques become commonplace in students' notes as they realise that, when they look back at a page of answers when faced with questions on the same topic, the answers are rather useless.

Testing with limited notes

I confess to stealing this idea from the Modern Foreign Languages department. You may not know that, these days, as part of the GCSE speaking assessment, students are allowed to take in with them a crib sheet of notes. This is limited to twenty words and as many pictures as they like. Consider setting this same restriction, or similar, for a halfway house between the open-book test and the formal testing mentioned above.

You could change the restriction, based on the topics you have covered. For a test on data handling that included methods of statistical sampling as well as averages from tables of data (groups and otherwise), I allowed: as many numerical examples as they wished; as many diagrams as they liked, but words on them counted towards their word limit, whereas words in tables didn't count; and thirty words beyond these restrictions. The results were remarkable and provided valuable rehearsal in 'making notes' and 'revision', the words that we so often see as homework for formal assessments and yet are meaningless skills to many students.

Blank-page testing

This is one that I use sporadically as a mid-unit starter activity. It is, in essence, an assessment, if a very broad one. I give students a blank piece of A4 paper and ask them to write down as much as they can on the topic(s) that we have been studying. This can include a worked example, step-by-step explanations, key terms and definitions or important formulae and their meaning. Be warned –

students need prior knowledge that this is going to happen, otherwise you will just get a blank look and a subsequent blank page. However, with warning and regular use, students get better at these, and they can be a really useful way for you to see what students have and haven't absorbed from the last few weeks of work.

Skills explanations testing

This is, in essence, a combined version of the testing with limited notes and a step more structured than the blank-page testing. This could be used as the assessment itself or in preparation for an assessment. You have guided the students into the key things they need to remember. They need to complete everything they think they need to be able to apply these skills successfully in a formal testing setting. The figure is taken from Year 11 revision preparation in December, and so covers a wide range of topics. As homework, students were asked to complete the grid. This was then marked and returned prior to the test itself.

12 key skills – key skills I need to remember		
Circles – area	Circles – circumference	Pythagoras
Volume of prism	Percentages – how to calculate a percentage of an amount	Indices – how do I simplify two powers x together
Expand a pair of brackets	Calculate a median average	What does it mean to translate a shape?
Solve linear equations	Straight line graphs – plotting and using	Fractions – how do I add two fractions?

Diagnostic testing

I dithered for some time about whether to even include this aspect of my classroom routine in this section of the book, because, at its heart, it is not about testing for recording; it is totally about assessment for learning and, indeed, teacher learning.

The greatest thing that I felt I learned in my first few years of teaching was how students go wrong. Generally speaking, as Mathematics teachers, we are relatively confident mathematicians. I know that may not be the case for those teaching Mathematics as a non-specialist, but even they would be expected to be able to explain the technique they are teaching to students in a clear and methodical way. What we cannot be prepared for, until we experience it for that first time, is how they get things wrong.

Knowing the common mistakes, how students perceive different elements of an explanation and how they then go wrong comes from experience at the chalkface. You can, however, work with your colleagues and your students to consider what mistakes they might make. For this, I use multiple-choice questions. Back in the Dark Ages, multiple-choice questions used to form part of the old O level assessment. As a final test, this is the worst kind of assessment. The answers were not merely random solutions but, rather, all common mistakes for that question. Hence, with only one mark for each correct answer, but, generally speaking, three or four substeps required for the solution, you could do three out of four correctly, but make a mistake with the multiplication of a negative, and bingo – nil points!

The magic of the use of diagnostic questions in the classroom is that they can help you to immediately identify what students are getting wrong and then address this. A quick search on the Internet for 'diagnostic maths questions' will bring up many resources that you can use. I don't mean to sound a geek, but I have always made my own. When I'm working with colleagues, diagnostic maths questions generate powerful discussions with a group of experienced and inexperienced staff as to what the 'wrong' answers should be.

ACTIVITY: DIAGNOSTIC QUESTIONS

For these 3 created 4 answers – 1 correct, 3 incorrect with a common misconception or mistake:

Pythagoras

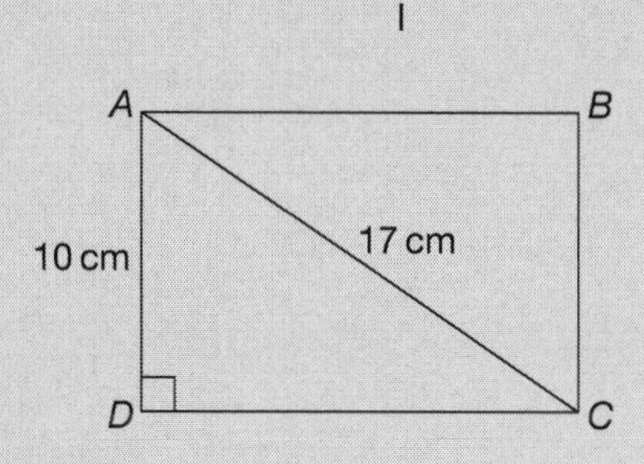

Diagram NOT accurately drawn

ABCD is a rectangle
AC = 17 cm
AD = 10 cm
Calculate the length of the side *CD*
Give your answer correct to one decimal place.

Compound interest

Ben bought a car for £12,000

Each year the value of the car depreciated by 10%.

Work out the value of the car two years after he bought it.

Fractions

Work out $\frac{2}{3} + \frac{1}{5}$

Have a think for yourself about what four answers you would give, before looking at mine.

Pythagoras

My solutions were:

- 7.0 cm: the student has remembered that the values must be subtracted, but has completely forgotten the squared aspect of Pythagoras.
- 13.7 cm: the correct answer.
- 19.7 cm: the use of squares has been recalled, but the student has not considered the hypotenuse aspect of the rule and has added the two squares before rooting.
- 189.0 cm: on the way to the correct solution, but the root at the end has been forgotten in the excitement.

Compound interest

My solutions were:

- 9,600: the student has calculated 10% correctly and understood the two years, but has used a simple interest calculation rather than compound.
- 10,800: a 10% decrease has been applied, but just the once.

- 1,200: the student is able to calculate 10%, but no decrease has been applied.
- 9,720: the correct solution.

Fractions

My solutions were:

- $\frac{3}{8}$: add the numerator and denominator. (As an aside, would you use those words, and, if not, why not?)
- $\frac{13}{15}$: the correct answer.
- $\frac{3}{15}$: the student has correctly remember the need for a common denominator, but has then merely added the numerators.
- $\frac{11}{15}$: a rather quirky but not uncommon mistake that is almost there. The numerator is correct; the cross-multiplication aspect of the numerator has gone wrong. The student has multiplied by the denominator of its own fraction rather than the multiplier needed to change to the denominator.

Clearly, there are many answers that you could give the students to pick from. The next step may be to give them such questions and the four answers and ask them to explain what mistake has been made, or, even better still, get them to make up the answers by using one single mistake in their method.

The importance of feedback

I mentioned feedback earlier when I talked about personalised assessments. There are lots of tricks to make this simple, but here are ten that I use or have used. They might apply to after a formal assessment or test or just as part of your routine marking of homework or classwork.

WWW and EBI

Many teachers have something like this on a stamp or use stickers. After each assessment, you write something in each box. The consistency focuses students on the formative nature of your feedback and creates an immediate target for them to improve upon.

What went well	*Even better if*

PAR

This is very similar to the WWW and EBI idea, but I like PAR as it explicitly involves the student in a dialogue about their learning. PAR – praise, action, response – forces you to have three stages, rather than just the two in the WWW and EBI model. The first is some positive feedback, the second is an area in which to improve and the all-important third is an expectation that the student will now implement something to improve this. This might be that you expect the student to comment on whether they now understand your action point or something more precise.

Question Time

This is another one that I use. The table is printed on stickers and, after marking students' books, I stick one of these in and add my comment and a question for them to do. The first box contains a comment on the work, probably something positive and an improvement. The question is then bespoke for the student (probably from a selection of five or six across the class) on something that they need to improve upon. We start the lesson with them reading their feedback and answering their own question, having read my feedback. We then go through all six questions together as a class. The feedback is personalised, and they can see immediate progress from getting the question right, after feedback (well, with any luck!).

Miss Upton comment	***Your answer***
Miss Upton question	

DIRTy time

You must then give time for them to answer the question. Alex Quigley[2] would call this DIRTy (dedicated improvement and reflection time). This might mean that, for the following homework, students are asked to respond to the question posed, or perhaps at the start of the lesson when the homework is returned this creates a focus to feedback in a very personalised way. It doesn't take long, later within the lesson, to check how the student did with the question you posed. Quigley suggests five tips for making this DIRTy time powerful and purposeful. My Question Time example ticks all the boxes for four of these:

1 Keep it focused. Simply handing back work to students and asking them to read it, look at their mistakes and think about them will be time wasted. It is not that teenagers are inherently lazy, but that they simply need coaching and a structure in order

to reflect upon their work. Those who have got all the answers right might think that this is a time to sit back and relax, whereas, with a focused question to consider, even those who achieved the highest scores can be pushed to the next stage of their reasoning.

For the student who has got much wrong, there is too much for them to take in. The clear focus of a single question allows them to address one point at a time. Come back to other issues later.

2. Model and scaffold. You might spend the first few minutes doing some examples on the board that mirror those that you have set in your Question Time. Tell students to focus on what you are doing and see how this applies to the question they have been set. They are then immediately listening, with a clear purpose, rather than the rather drifting attention that you often get during assessment feedback.
3. Targeted feedback. The precision of Question Time means that you are targeting a specific area of the topic covered or a particular error made by that one student. It doesn't take students long to recognise that they don't have the same question as the person they are sitting next to, and that this means it is important to them. The power of personalisation is an increased focus, as they know it is in their, rather than a global class, interest.
4. Exploit the power of peers. In Question Time, there is a simple way of reminding students to use their peers before asking the expert. Given that the likelihood is that, in a group of four students seated near each other, there are at least two different questions to be tackled, they may well have been successful in the first instance in a skill that someone else then needs to practise. It is easy to spend 2 minutes before you go through the solutions asking students to look at each other's corrections and see if they can ensure their accuracy before you go through them as a class.

Clearly, there is further scope for the use of students in the explanation of the solutions to questions posed. All you need are the names of five students in your markbook who got the types of question that you have posed correct, to celebrate the success of students and to give them the responsibility of teaching their peers as you go through the feedback.

Skill focus

This again I would stick into a book as students get a piece of assessed work back and, together, students would consider what they did well on and what they need to do to improve. Students highlight the skills and then make a comment on how they will work on the ones to improve.

ASSESSMENT: LINEAR GRAPHS

My mark: 82%

I did well on:

☑ plotting linear graphs from a table;
☑ identifying the gradient from a given graph;
☑ identifying the y-intercept from a given graph;
☑ drawing the y-intercept from the equation;
☑ drawing the gradient from the equation;
☑ dealing with equations that were not in the form $y = \ldots$

I need to go back over:

☑ plotting linear graphs from a table;
☑ identifying the gradient from a given graph;
☑ identifying the y-intercept from a given graph;
☑ drawing the y-intercept from the equation;

- drawing the gradient from the equation;
- dealing with equations that were not in the form $y = \ldots$

My tip for the future:

- *I need to remember that the gradient is the number attached to the x.*
- *When the equation is not y = ... I must rearrange it so that it is.*

  2y = 4x − 10 (divide each side by 2)

  y = 2x − 5

Of course, the quality of the comments varies, but this can be an area that you focus your attention on as you wander around the class. The more you do of this type of reflection, the better they will become, over time.

Traffic lights

After a lengthier assessment that covers a range of skills, I will often use some form of traffic-light sheet to focus students' reflection upon the assessment. This personalises the feedback, but involves the students in that process. You might initially train students in the completion of these by writing them in yourself, before they have to complete them after coverage of a topic or set of topics. You could equally do them before and after an assessment to focus their revision.

Skill	*Red*	*Amber*	*Green*
Drawing inequalities on a number line			
Simplifying (solving) inequalities			
Simplifying fractions			
Combining fractions (numbers)			
Combining fractions (algebra)			
Working out percentages			
Doing percentage increase/decrease			
Doing reverse percentages			
Doing compound percentages (interest)			
Drawing inequality graphs (simple)			
Drawing inequality graphs (complex with y and x)			
Designing surveys			
Types of selection for a sample			
Stratified selection (in groups)			

Oral feedback

My fifth type of feedback is the fifth aspect of Quigley's DIRTy time. Consider how you might capture ad hoc and oral feedback. I have a confession that many teachers will loathe me for. I hate sitting down in a lesson. Unless the class has been instructed to work in silence (only ever the norm in a formal assessment setting in my classroom; I just can't stand the silence otherwise), you won't find me seated in my chair at the front. I constantly patrol the room and, what is probably rather annoying, constantly look over the shoulders of my students to give instant feedback and see how they are progressing. This means that I am constantly posing questions about their work and making suggestions on improvements.

How do you capture this? Or get the students to recognise that these are the kinds of question they should be posing themselves? One quick and easy way is with a feedback stamp. There are many sites where you can order these, at little cost. Simply carry one with you as you wander and put the stamp in the book (probably next to some guided methods, as you explain where the student has gone wrong) to indicate to you and the student that they have received help here.

It might also focus their minds when you set homework or ask them to review their work. This type of ad hoc feedback is ever-present in the best Mathematics lessons that I observe. Consider how you will do this and how you will acknowledge, with the students, when, where and for what this feedback was given.

Frequency of assessment

How often should assessment take place? Interestingly, in Shanghai, students have SAT-style assessments every 3 months (as opposed to our Year 6, Year 9 and Year 11 formal external-assessment structure). I would not advocate that this is necessary or desirable, but I would advocate that frequent feedback on written work, through regular marking of classwork, homework or more formal assessment, is absolutely necessary in the Mathematics classroom.

I would mark classwork and homework weekly and set more formal assessments every half-term. These assessments, as previously discussed, can take different forms in order to see how students work in different contexts and with types of question.

At the heart of all good assessment is good feedback. Students value this and respond positively to it. The best teacher I know marks all her students' work, classwork, homework and formal assessments, and her students love her. She gives praise where it is due and, in books, acknowledges contributions in class and excellent effort. Your marking shows the students that you care about what they are producing, and, in turn, they care more about what they produce and take a greater pride in their work. Why would you care if no one is going to look at it?

TALKING POINTS

- What do you want to achieve from assessment?
- What forms could assessment take for different purposes?
- Are there expectations on assessment within the school?
- How will you give feedback, and how often?
- How can you give time for students to reflect upon feedback within the lesson?
- How can you scaffold that reflection so that they build skills?
- How will you create a personalised dialogue about learning with each student?

Your thoughts

Notes

1 C. Duhigg (2012) *The Power of Habit: Why we do what we do in life and business*. Toronto, Canada: Doubleday.
2 See: www.huntingenglish.com

6 Differentiation

In a subject where the answer to a question on differentiation is often 'we teach in sets', this chapter will consider why differentiation is essential in the Mathematics classroom, and how this is achieved, when ultimately all students must understand how to use each concept taught. When outcomes of questions are absolute, how do you teach a variety of abilities and students who will inevitably work at different rates? It will consider how to stretch the ablest and support weaker students, and how lesson planning should consider every individual in the classroom. This doesn't mean planning ten lessons in one, but will give practical advice on how not to teach in a 'one size fits all' way.

What does differentiation mean?

In Chapter 2, I asked you these questions with regard to differentiation, as a consideration of the elements of outstanding teaching from the HoT learning pyramid. Consider them again, as we focus on differentiation and what it might look like in the Mathematics classroom:

- Does your lesson have appropriate resources for those students who learn and understand conceptually, but struggle to write complete notes?

- Does your lesson allow for those students who can listen to your explanation and fully understand what to do, as against those who need to see the problem visually?
- Does your lesson consider completion of a range of questions, of increasing complexity, and will *all* students have to complete *all* questions?
- Does your lesson explicitly teach the skills of mathematical presentation, or does it make the assumption that all students will just do as you do, following a single worked example?
- Does your lesson allow for those students who were absent last week and missed a key explanation or concept?

Seven steps to differentiation heaven!

I will give some examples of differentiated activities in the Mathematics classroom but will first mention some aspects of differentiation. I call them my seven steps to differentiation heaven!

Step 1: accept that students are different

You have a responsibility to plan learning that suits the needs of each student. Students aren't all the same, and you must be open to different approaches and strategies that allow students to shine. Students want to be able to be original, resourceful and active. Consider ways that your teaching can allow them to do this.

Step 2: have variety in your lessons

Some students like working quietly on their own; others like the support of group work. Some like exploration activities where they have to discover and draw conclusions; others like to be told the right method and employ it in a structured way. I have deliberately avoided calling this spoon-feeding, as I don't think these students lack responsibility, but they struggle with more open tasks. Make sure that your lessons offer a variety of activities that allow different students to shine in different situations.

Step 3: use assessment to inform differentiation

Knowing your students is key to creating differentiated approaches. Only by knowing which students constantly struggle with negatives, or which struggle to visualise patterns, will you be able to support them and plan appropriately.

Step 4: be prepared to explain things in different ways

I think one of the hardest things when you start out in teaching is not realising that some students still won't understand, even after your clearest explanation. As a confident mathematician, if someone explains something to you, you tend to get it. For the students who struggle this is not the case. Be comfortable using different approaches, explaining things in different ways and using diagrams and visual representations to show abstract algebraic or numerical concepts.

Step 5: offer choices in the classroom

Where possible, give students choices as to what they do and how they do it. This might include offering a range of questions to complete, using a computer or calculator to support work on graphs, drawing diagrams alongside written working or showing working in a particular way. Students will work at different speeds and have different comfort levels of thinking abstractly, so offer choices to help make the abstract concrete.

Step 6: group students differently for different purposes

Don't let your seating and grouping of students in your classroom be static. It is all too easy to fall into the trap of creating mixed groups or seating arrangements, so that the most able support others and bring them along with them. Although there will be value and worth in those students learning to explain their methods to others, don't get stuck in a rut with how you seat students. Sometimes put them in ability groups rather than mixed-ability groups; this makes it much easier to differentiate the task and set different outcomes for different groups. If they are seated in groups, it also makes it easier for you to get around and speak to the students. With a class of thirty students, in an hour lesson they each get 2 minutes of your time (in theory!), but six groups of five students would each get 10 minutes of your time.

Step 7: realise that great teaching takes time

Great teachers are not born overnight. Knowing the nuances of how students learn, finding five different ways of explaining the same thing, all take time and experience. Think about one lesson and area of Mathematics at a time. Don't try something new every lesson; instead, take stock of a few lessons each week and think carefully about their activities and planning.

ACTIVITY: WHAT DOES DIFFERENTIATION ACTUALLY LOOK LIKE?

What does this mean in reality? Let's take a topic: straight-line graphs.

You want to teach the concept of $y = mx + c$.

- Where might you start?
- What activities might you do?

- What alternatives would you offer?
- How might you group students to explore or consolidate within this topic?

Have a moment to think about it.

Here is how I would teach this topic (be warned: I don't think I have ever taught the same lesson twice; I constantly vary my approach, dependent upon the group, and so this lesson might not work off the shelf for a group that you teach, but hopefully it will give you food for thought). The figure shows the mind map that I created to think about what I would include in the teaching of this topic and how that relates to our topic in this chapter – differentiation.

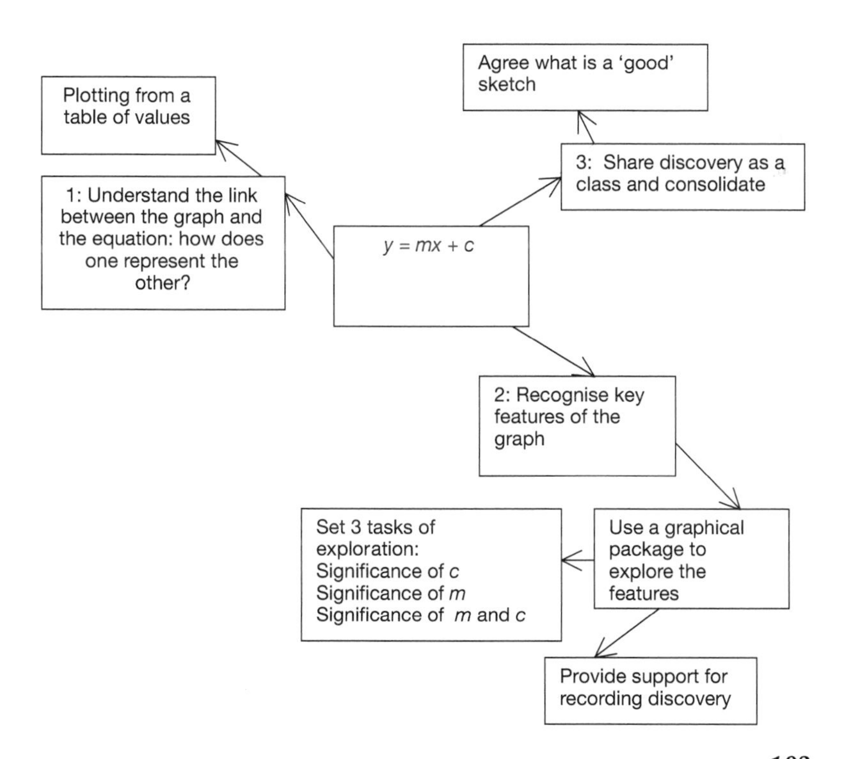

These might be my initial thoughts when I think about how I will teach straight-line graphs.

Stage 1

This will include some initial consolidation work on plotting graphs. This assumes that students can plot coordinates in all four quadrants and substitute into a formula, though both of these will be key points to mention as we start this work.

Critically, here I want students to develop an understanding of how the graph represents the equation. We will use a range of x values, probably −3–5, and we will talk about the use of a table of values to draw the graph. I am not expecting tables of values to be a new feature, but will give guidance on how to complete these, including consideration of which values are harder to work out: $x = -3$ or $x = 5$.

To differentiate, once we have gone through the key principles, I will give students a number of graphs to plot. This will include examples such as the following:

- $y = 2x + 4$ (a simple, all-positive example);
- $y = 3x - 5$ (subtraction is involved, but substitution still uses just positive coefficients);
- $y = 10 - x$ (will need careful thought for the substitution of the negative values);
- $y = -2x + 5$ (a different presentation but a more complex equation to substitute).

I might have ten equations to complete, but would group them into easy, medium and hard levels of difficulty, and I would tell students that they must do at least one question from each group. If they wish to focus on the more difficult, then they may do so.

I am differentiating the task and adding a layer of expectation of all doing some harder questions, but those who are more confident may tackle only the harder questions. After 5 minutes of working on these, I would 'float' around the class and make sure that students have made an appropriate choice for their ability level.

Stage 2

I would use a graphical package, either on the computers or a graphics calculator, in order to get students to see and consider as many examples as possible as they explore the meaning of $y = mx + c$.

I would give them a worksheet to structure their recording of the graphs that they see, something like the figure.

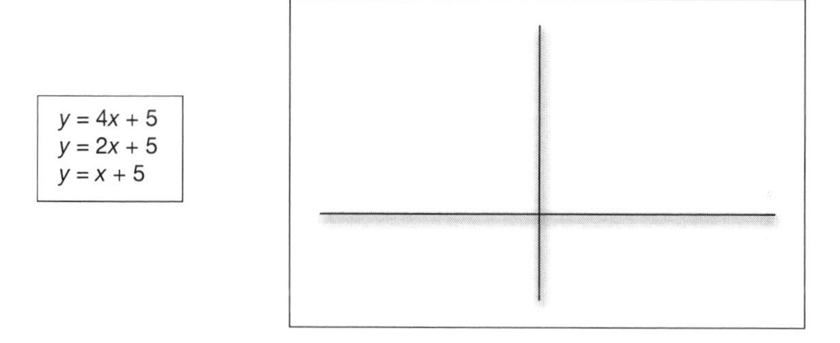

Type in the equations and then draw the graphs in the box on the right. Highlight each graph to distinguish which equation it matches.

I would create three tasks, one that focuses on the significance of c, one that focuses on the significance of m, and one that has both. I would assign the students a task based on their ability, the most able doing both m and c. I would make the decisions on who did what. Students working on the same task would be seated together to share their outcomes. Students would have Task 1, 2 or 3.

After exploration (note the expected prompt of 'I noticed that . . .'), we would come back together, and I would change the seating. Students would then be grouped so that each group had students who had done Tasks 1, 2 and 3. They would be asked to look at their discoveries and together decide on a 'class rule' for the link between the equation and the graph.

Throughout these activities, I have been considering who works with whom, creating variety in this and greater challenge for abler students. I have given a structure in the format of a worksheet, which had more structure for those who needed more support. In the original graph-plotting activity, students have chosen the level of difficulty; in the next exercise, I have chosen those groupings.

Stage 3

We finally come together, and each group of three has to explain its 'class rule'. The students have been given one side of A4 to do this, with marker pens. This activity appeals to the more creative and allows me to reinforce what good notes might look like in our books, in order to refer to this later. We decide as a class what we want our notes to include, and then all copy into their books.

The next step would be to consolidate this with some worked examples. I would do this at first using mini-whiteboards. This gives pace to the work and allows those who are a little unsure still to get instant feedback and be able to 'have a go'. We do five examples of 'sketch the graph of this equation' to ensure that all have a grasp of the significance of m and c.

Next, I would either set a sort/match activity of some graphs and equations and ask students to match them up, or they would sketch some graphs from equations given, formally in their books. The equations to sketch could be of an increasing level of difficulty and would include some at the end where y is not the subject. This would then push the more able to consider what happens in this case.

If possible, when the students are completing these sketches in their books, I would still allow access to the graphical package for them to check their solutions. This would mean that students could be responsible for knowing if their work was correct and could continue to use the graphical package for support and variety as needed.

Seven steps to differentiation heaven – applied to our problem

So, how does my planning of straight-line graphs stack up against my seven steps to differentiation heaven?

Step 1: accept that students are different

Different activities and responsibility for use of computers and resources at different points allow students to make appropriate choices.

Step 2: have variety in your lessons

The sequence of lessons includes a variety of exploration and consolidation, individual and group work, activity matching and straightforward questions.

Step 3: use assessment to inform differentiation

The sequence of lessons starts by considering what we will need to recap and who might struggle and, in the plotting exercise, allows students to enter at their own comfort level, while having a minimum completion for all.

Step 4: be prepared to explain things in different ways

Although graphs form an easy area to show visually, it would have been easier to avoid use of the graphical package, and, unfortunately, many Maths teachers do, other than for whole-class instruction. Giving students opportunities to work in different ways, even if the final exam that they need to pass will be paper-based, creates a variety in your lessons that will engage and motivate students.

Step 5: offer choices in the classroom

In the later stages, the computer package will still be available for students to check their working.

Step 6: group students differently for different purposes

The grouping is fluid throughout the topic. We learn different things from working with different people. The tasks are set to differentiate, and yet mixed groupings are used to share findings. This is all under the assumption that this group is a set, but recognises that students within that group have different abilities.

Step 7: realise that great teaching takes times

My planning starts with an overview of what I will do and considers the resources, groupings and activities that will take place within it. The individual-lesson planning comes later and will start to address what the prerequisite knowledge is and the examples that will be used to instruct the class.

So, differentiation in Mathematics *is* possible and *is* necessary. Consider how you will do it and make your methods varied and personal.

TALKING POINTS

- How will you make sure that you differentiate in your classroom?
- What does differentiation mean in Mathematics?
- Take one topic area and think about the seven steps to differentiation heaven and how you would implement them. Try it for, say, volumes of prisms.
- Think about the way that you explain things: Why does it work that way? What are you doing at each stage in a problem, and why? How did you break down that solution?

Your thoughts

7 The importance of literacy

Language and literacy are no less essential in the Mathematics classroom. In this chapter, consideration will be given to numerical literacy, lexical literacy and the language of explanation (how to make the impossible sound possible), and clear examples of how you can make complex concepts clear, without dumbing them down. Literacy in Mathematics has to move beyond keywords and definitions. In recent years, there has been an increased emphasis on literacy in all subjects; this is absolutely correct and necessary, but this has been taken on board in most Mathematics classrooms in a rather tokenistic manner. In this chapter, we will get into what mathematical literacy means and the value of this for mathematical fluency.

Precision of language

The first most important advice that I can give for use of language in the Mathematics classroom is *precision*. If there is a formal mathematical term for something, don't avoid its use in preference for some alternative articulation. Let's take reflection as an example. Many students learn at a primary level that, when a shape reflects, it 'flips over'. The number of times that I have seen this response on GCSE papers is really worrying. With this response, the student clearly understands the process that is taking place and can recognise the visual geometry that has been performed. They will, however, get nothing for their description, the only mark being awarded for the precise term, reflection.

Consider the four basic numerical operations and the words that are associated with them:

- addition: plus, increase, total, sum;
- subtraction: take away, minus, less, difference, decrease, deduct;
- multiplication: product, times, lots of;
- divide: share, quotient, goes into, how many times.

Our primary colleagues have been working hard to improve the use of accurate technical terms with younger students, but there is still much overuse of the words 'times' and 'share'. Our counterparts in the Far East are religious in their precise use of language: they don't dumb down words such as 'denominator' and 'factor', and students learn the appropriate technical term from their first encounter.

In this chapter, I will consider different aspects of how we can make literacy part of the Mathematics classroom. I will look at:

- activities that bring literacy into the classroom: bolt-on activities that you might use as a starter or part of a lesson to reinforce literacy;
- the language of explanation: here, I will look at how important spoken language and high-quality talk are within the classroom, and how you might formalise these;
- the importance of reading: how much time you spend considering how students read and interpret questions, and why this is something we should do explicitly;
- teaching of variation: here, we will think about how, as teachers, we develop our questioning in its mathematical language to develop understanding and to teach increasingly complex concepts;
- written solutions: of course, as part of any mathematical literacy, we must consider the language of mathematical

solutions and how students express them. How much do we care about how they write their methods and solutions?

Activities to bring literacy into the Mathematics classroom

Heard the word

Use this activity to reinforce key terms, perhaps as a starter. Are the students clear on the definition of each word and what it means? The table is given on a piece of paper and stuck into their books. This short word group followed on from a unit of algebraic work, but was given as a starter a couple of months later.

Word	*Never heard it before*	*Have heard it but don't remember what it means*	*Have heard it and it means . . .*
Equation			
Expression			
Variable			
Gradient			
Substitute			

Mathematical concept wall

I confess to stealing this idea from Mr Collins,[1] a Mathematics teacher who writes a blog on his activities in the classroom. I have used it in my Mathematics lesson and created a wall display from it, which we then come back to for regular tests and literacy reminders.

Give students a card with a variety of words on, such as these (I used this one after a range of work on shape):

volume	prism	net	units
interior	circumference	pi	radius
cube	cuboid	net	rectilinear
measure	cylinder	surface area	perimeter
pyramid	quadrilateral	circle	construction
scale factor	vertical height	skew	sector
segment	chord	tangent	triangle
bisector	cone		

On the reverse of the card they find this:

> Word:
>
> Pick a word, or two from those on the reverse of this card and draw or write a definition to explain what they are. Try to link any mathematical concept to the word(s) that you have chosen.

Mathematical activity cards

Create mathematical word cards, laminate them and have them available for a range of activities. This is an example set, but you can make them on anything.

Area	Perimeter	Volume	Fraction	Decimal
Percentage	Ratio	Square number	Average	Multiple
Factor	Solving equations	Substitution	Integer	Prism
Rounding	Reflection	Rotation	Enlargement	Tessellation
Polygon	Square root	Inequality	Plan	Prime
Estimation	Inverse	BIDMAS	Sequence	Parallel
Angle	Equivalent	Graph	Units	Convert

Possible activities include the following:

- Give students a card each and ask them to group themselves with other people who have cards that somehow link to theirs. Ask them to explain to the class why they think they go together. Discuss if this is the only way they can organise themselves, or does anyone think they can move groups?
- Give students a card/cards at random. They can then make up a question based on the topic on their card.

- In pairs, give students 30 seconds each to describe their word to their partner for them to guess – they can't say the word or any part of it though!
- Splat! Put the class into teams and stick some or all of the words to the board. Ask the teams to number each member, for example, from 1 to 5. Player 1 from each team then comes out. Either the teacher or another learner describes the word, and the student who 'splats' it first with their hand gets a point. You can remove them from the board as you go along or leave them all up. All Player 2s then come out, and the game continues.
- Just like Pictionary, students have to quick draw their word for others to guess.

These activities all work to promote the use of appropriate language in the classroom and to reinforce key concepts and terms. However, these types of activity alone will not create a language of Mathematics in your classroom. There must be more depth than this.

The language of explanation

How exacting are you going to be about students' spoken language in the classroom? Will you give them a structure to go about this in the best way, and how can you insist on precise vocabulary? There has been considerable research that shows that a focus on high-quality talk in the classroom results in better-written work. This is true of the Mathematics classroom as much as it is the English or History classroom.

We must consider the language that students use when they describe techniques. Debra Myhill, although an English specialist, gives many examples of why this is important. In her paper 'Talk, talk, talk',[2] she explores the following questions:

- How do teachers use talk in whole-class episodes to scaffold learning and develop understanding?

- How interactive are whole-class episodes?
- How do teachers build on prior pupil knowledge?
- How do teachers use questions?
- How do pupils use questions?
- How is the handover to independence or critical moments handled?
- What do teachers believe about talk as a tool for learning?

One Ofsted report, 'Mathematics – understanding the score',[3] is very clear that high-quality talk in the classroom is essential for students to learn and make progress. The report, which was based on evidence from 192 inspections in maintained schools, from April 2005 to December 2007, strongly advocates that talk enhances the learning of Mathematics when pupils have regular opportunities to:

- describe, explain and justify their understanding of mathematical concepts;
- practise using precise mathematical vocabulary;
- 'think together', discuss, exchange, explore, develop and revise ideas with each other;
- share their mathematical reasoning and understanding;
- communicate face to face with an audience;
- rehearse ideas, their reasoning and language structures before writing.

One way to make this spoken language more formal is through writing frames and prompts such as these:

- I found it helpful to use . . .
- I knew that . . . wasn't correct because . . .
- I noticed that . . .
- At first . . .

- I know that . . . because . . .
- One problem was . . .
- Stages that I have used to solve the problem include . . .
- I think the answer will be . . .
- I think my solution is correct because . . .
- A possible solution would be . . .
- An alternative method would be . . .
- Checking my method has shown . . .

Oral frames are a strategy for structuring student talk and for impressing on them the importance of formal use of language in a classroom setting.

The importance of reading

How often do you consider how students approach the language within a question, and how much time do they spend decoding what the question is asking? How many times have teachers asked students to 'Read the question carefully', without really explaining what this means. We must explicitly spend time with students helping them to decode questions and fully interpret the meaning of text and mathematical expressions.

Thinking about language in the Mathematics classroom should also consider how students interpret questions. A head of Science, whose department was top of the league tables, once told me about how he had been asked to show an inspector what he did in Science that secured such impressive results. His conclusion was that students read for understanding, spoke with purpose and wrote with a precision that demonstrated their knowledge and understanding. Underlying this success, he felt, was the excellent teaching and learning that took place within English in the same school. Of course, this anecdotal story implies that it is all up to the English teachers, which I do not believe. It does, however, reinforce that success will be linked to the precision of language used in the classroom and the confidence with which students use this in speech and, subsequently, in writing.

ACTIVITY: DECODING LANGUAGE

Here is one activity that I used with a GCSE class that was struggling with the interpretation of language in questions on percentages. After a couple of weeks of learning a range of techniques for calculator-based calculations with percentages, I had then given the class a range of problems with different needs and different contexts. The students were awful at them, not because they couldn't apply the mathematical technique, but because they could not infer from the question which technique they needed to apply. So I stepped back and gave them these twelve questions:

- A gentleman, after 2 months at Weight Watchers, loses 35% of his weight. When he started, he weighed 99 kg. How much weight did he lose?
- A stereo costs £320. In the sale, it is reduced by 25%. What is the sale price?
- A season ticket for Bolton FC this season is to go up by 20%. It will cost £788. How much was it last season?
- John throws the javelin 52 metres. His next throw is 12% better. How much further did he throw?
- Peter invests £500 in an account paying 1.4% interest per annum. How much will he have after 5 years?
- A town had a population of 26,200. The total has fallen by 8%. What is the new population?
- Gary Barlow's latest single sells 65% fewer copies this week than last week. It sold 29,750 copies this week. How many did it sell last week?

- Sarah buys a car for £25,000. The first year it depreciates by 25%. The next 2 years, it depreciates by a further 10%. How much is the car now worth?
- A shirt is slightly shop-soiled and so it is reduced by 30%. It was £40.50; how much is it to be sold for?
- A woman buys a house for £162,000. Ten years later she sells it for a 30% profit. How much more does she sell the house for?
- Ian buys a second-hand car. The price falls by 24%. He sells it for £2,356. How much did he pay for it?
- Fred borrows £4,000 on an interest rate of 5% over 1 year. After the first year, he has only repaid half of the money (including the interest). The interest rate then goes up to 7.5%. How much money will he owe after another year?

I then asked the class to group the questions into four categories; for further support, I gave some students an example of a type of question for each category. I stressed that, at this stage, I did *not* want them to answer the questions, just to decide which types of question they were and, hence, which mathematical technique would be needed to solve the problem.

For the brighter students, their four-categories sheet looked like this:

Normal percentage (Work out a percentage of an amount)	*Percentage increase or decrease*
Reverse percentages (Work out the original amount after an increase or decrease has taken place)	*Cumulative percentages* (Apply percentage change over time – simple or compound)

For those who needed some support, their four-categories sheet looked like this:

Normal percentage	*Percentage increase or decrease*
(Work out a percentage of an amount)	
Example: A shirt costs £50 but has a 20% reduction in the sale. How much is taken off the price?	Example: A shirt costs £50 but has a 20% reduction in the sale. How much is it now?
Reverse percentages	*Cumulative percentages*
(Work out the original amount after an increase or decrease has taken place)	(Apply percentage change over time – simple or compound)
Example: A shirt is £35 in the sale, after a 40% reduction. How much was it before the sale?	Example: A shirt was £79. In the sale it was reduced by 40%. It has then been reduced by a further 25% of the sale price. How much is it now?

The students completed the task, and then we looked at how they had made their decisions as to which technique was needed for each question. I asked them these three questions:

- How did you decide?
- What advice could you give someone who was struggling with this?
- Are there any keywords or language features that are true for each type of question?

Responses included consideration of tenses rather than just keywords and triggers. The following work that they completed on these skills was far superior, as they had explicitly been asked to consider the language of the questions.

Teaching with variation

An important aspect of teaching in Mathematics, I think, is *teaching with variation*. I mentioned in Chapter 2 the range of questions in most modern textbooks, from questions that one might say are repetitive, giving students time to practise the same skills over and over, with one small variation (generally in the numerals of the question), to questions that vary greatly in presentation and language to make the students consider how they might apply the new concept in a variety of settings.

Teaching with variation is something that experienced teachers do without thinking; it means to me: teaching with a range of examples during the explanation of a single mathematical concept. The range of examples that a teacher uses to introduce a topic will come from a bank of knowledge built up over a number of years and through use of a range of resources. For someone starting out (and even for us old lags), what we ask, how we ask it, and how it is scaffolded should be given specific thinking time in our planning. I could probably be talking about this under the heading of 'Questioning' as much as the heading of 'Literacy'. In essence, I am talking about the questions that you ask students in order to build their understanding. For me, this sits neatly under literacy as a use of language (lexical, numerical, diagrammatic or algebraic), as it involves careful thought as to how and what you are asking – in other words, precision of language. It is irrelevant if that language takes the form of words or equations; my precise use of it will support development of understanding in the classroom.

Every lesson should include at least three examples, with varying connotations or levels of difficulty. I might not deliver these questions all at once; more generally, they are interspersed throughout the lesson to build the complexity of the work. This teaching with variation can also be known as 'concept variation'. For example, imagine teaching the concept of factorising quadratics:

Can we factorise the following: $x^2 + 5x + 4$?

Once this has been explored, the following question is posed:

How about $x^2 - 5x + 4$?

Then, four further examples are given to tease out different aspects of this skill:

$x^2 + 3x - 4$

$x^2 - 3x - 4$

$x^2 - 4$

$x^2 + 4x + 4$

And:

$x^2 - 5x - 4$

You will note, of course, that this last example to explore does not have a factorised solution – use of strategic 'incorrect' questions can be useful to explain why this one *doesn't* work.

By giving examples that vary in conceptual difficulty, you challenge students to think and develop a clearer understanding of factorisation of a quadratic. This range of questions over a single topic is essential for building up conceptual understanding and depth of knowledge of the process being taught, and so it is vital, when you consider questions you have set for consolidation, either in class or at home. Do they sufficiently develop the range of skills and give this variation? Variation should provide students with opportunity for drill and regularity, while also giving breadth to ensure that they have mastered the range of variation needed within a given topic.

Through this idea of teaching with variation, we are explicitly thinking about the language of Mathematics that we are teaching. We are considering the changes we make to a question to make different teaching points and, in turn, develop a fuller and more meaningful understanding of a concept.

The presentation of solutions

I shall start this section with an activity.

ACTIVITY: A MULTI-STEP PROBLEM

I am going to give you a multi-step problem. The problem is from the higher tier at GCSE and would be probably worth 4 marks in total and be likely to be near the end of the paper. Without reading any further and without consideration of all I have said with regard to precision and exact use of language (mathematical or lexical), think how you would display your solution to this problem. There is space after the question for your working.

A ship leaves port and sails on a bearing of 050° before turning and travelling 3.5 km on a bearing of 100°, where it stops at a rig. At 18.20, the ship leaves the rig and and heads directly to the port, with a speed of 1.2 km/h. Find the time the ship returns to the port.

So, what did you write, and, importantly, how did it look? This is my version:

First, consider the literacy of your students' diagrams: notice how my diagram uses a ruler, gives angles that are, by the naked eye, approximate to those stated, and my diagram is clearly labelled, with notation for the port, P, and the rig, R. My diagram then adds any further angles that I can deduce from my knowledge of parallel lines (the North bearing lines). The angles of 50° and 110° are placed on the diagram first, from the given information. The 130° angle can then be deduced from the corresponding or supplementary angles rule, and then the 120° angle can be deduced from the angles at the point. These are all added to the diagram before any more complex calculations are attempted. I subsequently draw out a simpler version of the triangle, with labels for the vertices as A, B and C. I do this in order to be able to use the cosine rule in its original form.

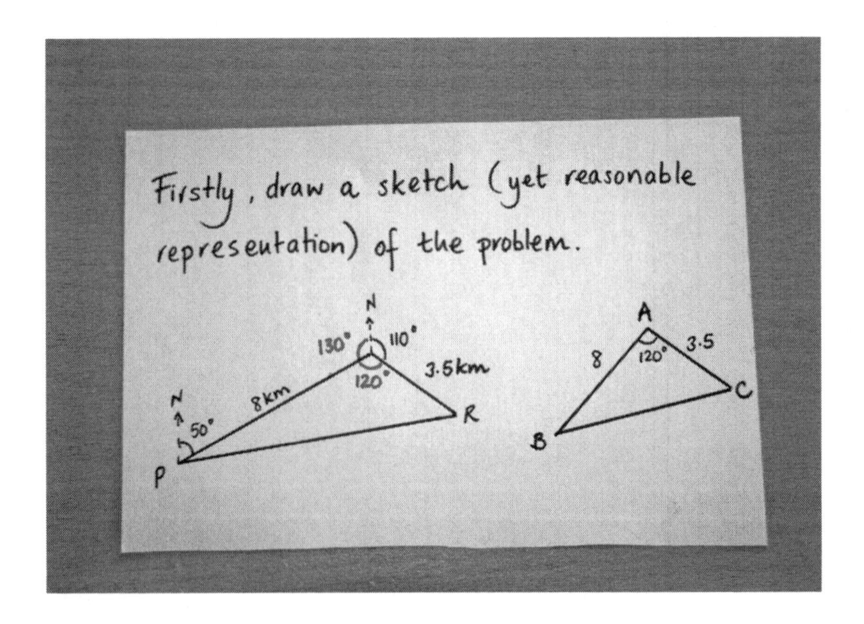

Second, consider the literacy of the written solution.

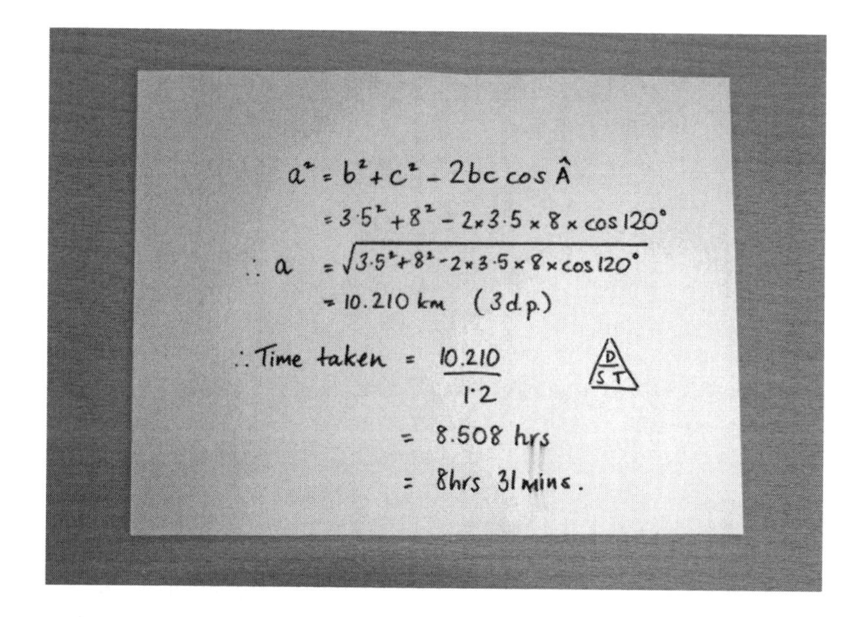

Notice how I have presented my solution. I have written out the formula to be used in full and substituted into the equation from my diagram. I have then shown the working needed to reach a solution for the distance travelled by the ship. After this, I have worked out the time taken, showing the speed, distance, time formula (and triangle to remember the rule) and then demonstrated the answer as a decimal and as hours and minutes. I have been clear about the degree of accuracy I have used for the reader (but at all times, of course, kept a full solution on my calculator).

- How did your response compare with mine? With no excuses for my messy writing, what do the presentation and the level of detail say about the expectation for numerical literacy in my lessons? How many of these expectations will be the norm in your classroom?
- Would you expect the sketch to be drawn with a ruler?

- Would you expect a further sketch with the A, B, C labels?
- Would you expect a student to put their equal signs underneath each other like this? Would you expect more working steps before the answer after the square root from the cosine-rule calculation?
- Would you expect the cosine rule to be written out in full?
- Would you expect students to state their rounding?
- Would you expect them to state the units on the cosine-rule solution?
- Would you expect them to jump straight to the hours and minutes answer or leave the answer as a decimal?

The emphasis on strict, formal solutions and a particular format to those solutions is important in your classroom. I talk to my students about how thinking can be messy, and how I don't always expect their working to be neat. However, these occasions are rare and often form part of a more open-ended problem, where I want them to work with others and consider different avenues. When solving a problem such as this, and many more, the presentation of the solution supports the logic of their thinking and the accuracy of the final answer, and, hence, it is absolutely essential.

That presentation is no less important when speaking. Imagine setting a task where the student has been asked to show and explain how they constructed the perpendicular bisector of a line. The student explains their method and does so reasonably correctly. However, their explanation is overly verbose and does not use specific mathematical language. Ask them to explain it again, this time using the smallest number of words. If this is not forthcoming, use others to develop the desired response and praise the use of elegant mathematical language. This requirement for precision is linked to the mathematical concept of proof, and, although students only come across this briefly at GCSE, it underpins the clear and logical argument required for true mathematical thinking.

So, the literacy of the Mathematics classroom is critically important, but don't think about it just in terms of keywords and definitions. Think about it as spoken language, accurate written solutions, the demonstration of logical thinking, the consideration of questioning and the decoding of language in questions – in fact, a lot more than just words.

TALKING POINTS

- How will literacy mean more than vocabulary and definitions in the Mathematics classroom?
- Think about being clear on the command words in questions and what they mean: solve, simplify, evaluate, explain, prove etc.
- How will you encourage, model or enforce good mathematical literacy with clear reasoning?
- How will you use talk to encourage good literacy (linguistic and mathematical)?
- Consider your language in the classroom and how this develops mathematical literacy and sets high expectations in the use of appropriate mathematical language.

Your thoughts

Notes

1 See: http://mrcollinsmaths.blogspot.co.uk/2013/01/mathematical-concepts-wall-for-want-of.html?m=1
2 D. Myhill (2006) 'Talk, talk, talk: Teaching and learning in whole class discourse', *Research Papers in Education*, 21, 1, 19–41.
3 See: http://webarchive.nationalarchives.gov.uk/20110809091832/http://ofsted.gov.uk/resources/mathematics-understanding-score

8 Questioning

One aspect of the Mathematics classroom that many take time to find their feet with is the way in which you phrase questions to students. I am not talking about the Mathematics questions that you set as an exercise. We have previously covered how it is important that these technical, contextual questions build the skill, with different nuances and elements, staging their understanding and, hence, building confidence. This time, I mean those broader questions that you pose to help a student develop their reasoning and unpick complex problems. Asking good questions and promoting discussion are integral parts of the best Mathematics classrooms. It is worth spending some time thinking about the broader questions that you will ask, and how this will shape the way that you make your classroom one where students and adults think.

There is a delicate balance to be struck with these types of question, as in fact they are often a response to when a student is stuck or when a student has finished a problem. To the student, this can be as frustrating as it is as a parent when your child goes through the 'Why?' phase in their learning: 'How does that work Daddy?', 'Why can't we see the stars in the daytime Mummy?' etc. Many a parent has reached exhausted frustration with the child who just goes on and on with one question after another. Of course, this is not something to discourage, but it can be sapping of your energy. The same feeling will occur with students if, as soon as they offer a solution, you say, 'How did you get that answer?' 'Are you sure it is correct?' 'How can you be sure?' These questions and more, as I will go on to discuss, are essential to build reasoning and a deeper understanding, but we must be careful that, as we question them, students are not frustrated with our constant barrage, when they seek reassurance.

I categorise my questions into three areas. They are:

1 questions for deeper understanding;
2 questions to clarify confusion;
3 questions to support when struggling.

In some ways, these could be related to the independent learning questions that we posed in Chapter 2, but these questions are specifically related to developing mathematical understanding.

Questions for deeper understanding

These questions would be used when you are working with a student who understands the problem and who may well have a correct solution. Your questioning is directed to challenge their methods and make them consider if these methods would work with similar problems. These questions can also be directed to get them to check the accuracy of their solutions. We have all been guilty of saying to students, 'If you finish your exam early, use your time to go back and check your working', but what does this really mean they should do? Use of these questions throughout your classroom teaching should mean that they can employ the questions themselves when in an assessment situation.

- Are you certain that is the correct answer? Why?
- How can you be sure that your method works?
- Would it work if I changed that to [a different number/term/shape in the original question]?

- Could you explain your method to someone else?
- Have you checked all your calculations (add, subtract, multiply and divide)?
- Where there are negatives, have you checked that you have dealt with them correctly?

Questions to clarify confusion

These are the questions that you use when a student thinks they are stuck, but, in fact, on looking at their working, you can see that they have approached the question in the right way, and they just need a prompt to get to the next stage in their thinking. Invariably in these situations, the student actually does know what to do next; they just lack the confidence to put their thoughts into action. In this stage of questioning, it is essential that your questions try to clarify their understanding thus far and help them to resolve the next step. You don't want to give them the answer, but equally you don't want to just look at their work and then say, 'Think about it'. There can be nothing more frustrating than this for the student. We want to use questions to build their self-reliance and ability to break down and solve problems for themselves, but we can't do this successfully in Mathematics without stepping-stones along the way. These questions need you to look carefully at the working that the student has done and assess what they need to do differently or next to complete their solution. This needs an understanding of what they have written, or explained orally, and a subtle approach that builds upon that.

- What have you done so far?
- What don't you understand now?
- Why do you think it is wrong?
- What is it about this question that makes it more difficult? How might we deal with that?

Questioning

- Could we try the problem with simpler numbers/terms and then apply it to this one?
- What do we know already that might help?
- Let's go back through your working and see at what stage you have gone wrong.

Questions to support when struggling

This is perhaps the hardest, and I can confess it can be very hard, the lower down the ability spectrum that you go. I didn't think that, when I started teaching, I would be considering ways of breaking down an explanation for how to subtract numbers from 100 to students aged 13. The numeracy skills of some students are very worrying; fortunately, these students aren't in the majority, but every now and then I stop and have to think deeply about how I can explain that any more simply!

These questions are what you might pose to those students who are faced with that mathematical brick wall and don't know what to do next. They are almost ready to throw their pen down in exasperation (if you are lucky enough to catch them beforehand), and you need to instil a confidence that will bring them back from that brink, and yet not simply answer the problem for them.

- Which words/numbers/terms are important in the question?
- What have we learned (recently) that might be useful?
- Have we come across other questions like this? How did we approach them?
- Where do you think we should start?
- What are we trying to find out?
- Can we draw something that will help? A diagram/graph/flowchart?

These three sets of questions could be something that you give to teaching assistants in your lessons to guide them in how they support students. This approach means that you and the teaching assistant are modelling the same support and are both utilising practice that you know will develop students' understanding of the topic that they are covering, but also their mathematical reasoning for all areas.

TALKING POINTS

- What questions will you ask to develop student understanding?
- How will your questions help students to develop their reasoning, rather than give them the answer?
- How will you strike the balance between questions that strip and build confidence?

Your thoughts

9 Ensuring good behaviour

Student behaviour will always be linked to the engagement and motivation in the Mathematics classroom. This chapter will think about the generic skills that a teacher employs to ensure good behaviour management and give consideration to this set in the context of the Mathematics classroom.

I once worked with a teacher who said that teaching was 95 per cent personality. I know what he meant, but I have to disagree. His implication was that those teachers who are full of life, interesting and always have something to say and have an engaging personality make the best teachers. I don't think this is true. I think that such teachers can often have a much easier time of classroom management, as they diffuse situations with humour and have a knack of saying things to students that make them smile and not misbehave. I don't think that these necessarily are the best teachers in schools. I could send you back to Chapter 3 on planning or Chapter 4 on assessment to remind you of the work that goes on behind the scenes that makes excellent teaching and learning, not just a winning smile and joke every lesson. All that said, classroom management and a well-behaved classroom are essential for learning to take place.

Doug Lemov, author of *Teach Like A Champion,*[1] gives some very helpful strategies for starting out in the classroom. The book is typically American in style, but with very transferable skills across the continents. One thing that is worth starting out with is his micro-technique for classroom management. Doug Lemov rightly says that students who are not concentrating will not make progress. He calls them leaky buckets. One acronym in the book to remind you of the importance of this is SLANT (see Chapter 5). Lemov says, 'No matter how great the lesson, if students aren't alert, sitting up, and actively listening, teaching them is like pouring water into a leaky bucket.' So it is critically important that, if you want them to understand when you explain about the construction of a perpendicular bisector, they are listening and attentive.

Generic classroom skills: micro-skills

During time teaching in one school, I worked with Geoff Barton on the micro-skills of a good teacher. It came about from a discussion with some aspiring middle leaders on how you would help someone who was not a teacher to 'fake it' in the classroom. What are the things that teachers with natural classroom management do that you could tell someone who was starting out? We are too often quoted as saying someone is a natural at that, but there must be things that can be taught. Geoff and I thought that there were, and we called them our micro-skills. These are the things that effective teachers do, without thinking.

Where the effective teacher stands

The effective teacher:

- circulates the room while the students are working and even while he/she is teaching;
- avoids putting a barrier between him-/herself and the class (would very rarely be found behind the desk, either talking or even seated);
- may deliver parts of the lesson from the back of the room, expecting students to look at the board and yet listen to what the teacher is saying;

- addresses the class from a position of authority – he/she might stand at the board or bend to talk to an individual, but always looks in control and has an awareness of the whole room;
- avoids no-go zones – doesn't get stuck in a corner of the room and has to make a meal of getting out that causes laughter in the class.

How the effective teacher stands

The effective teacher:

- when talking, stands still to allow students to clearly hear what is being said;
- waits until the class is silent before speaking and is confident enough to do this standing at the front of the class.
- moves around while speaking, but always pauses to speak.

Use of body language

The effective teacher:

- smiles, frowns, stares as necessary (there is much that can be diffused with a teacher look);
- uses commanding gestures and clear arm movements, points for clarity and gestures for emphasis, but these flourishes are planned and part of the act, not an involuntary habit that distracts from the point being made.

Use of eye contact

The effective teacher:

- sweeps the class;
- uses a glance to stop someone from losing concentration;

- looks students in the eye (but not in a staring, creepy way!);
- avoids having papers or books as a barrier when talking; if reading from a text, he/she brings the head up frequently and makes continued, regular eye contact with the class.

Use of voice

The effective teacher:

- waits before speaking;
- has a good volume for normal class delivery that is neither too loud nor too quiet (a normal volume should mean students need to be quiet to hear what you are saying; don't bellow all the time);
- uses an increase or decrease in volume for effect, to grab attention or to create a moment of atmosphere in the room;
- uses tone, pitch and texture to make his/her voice varied rather than monotone;
- avoids fillers such as 'okay', 'all right', 'know what I mean' etc.;
- says 'thank you' rather than 'please' with instructions;
- doesn't talk too much and uses students to speak to vary the input;
- uses students' names a lot;
- has alternatives to asking questions.

Use of silence

The effective teacher:

- pauses longer than a class might expect;
- gives thinking time after asking a question;
- waits for and insists on silence at key points.

Use of praise and feedback

The effective teacher:

- uses praise and feedback in a meaningful way and makes it personalised;
- has varied ways of giving praise – 'great response', 'clever idea', 'excellent reasoning', 'quick thinking', 'nice one', 'fabulous', 'I hadn't thought of that'.

Management of social dynamics in the classroom (the hardest to achieve)

The effective teacher:

- greets students as they enter the room – 'Welcome to my world!';
- makes jokes but does not try to entertain;
- does not use sarcasm to put students down;
- isn't personal about students in that class nor talks about students in other classes;
- doesn't court popularity with students;
- is prepared to stop the lesson and start afresh if the class really is lost;
- knows when the group needs to work in silence to settle and when it is appropriate to build a buzz.

Clarity of explanations

The effective teacher:

- explains in a clear but non-patronising way;
- allows students to ask questions and does not feel threatened by these;

- answers rather than avoids questions, even if the answer is sometimes 'I'll come back to that. It is a great question and will be relevant later.'

Following things up

The effective teacher:

- says what he/she means;
- insists on silence when it has been asked for;
- ensures that, if the punishment stated for late homework is detention, this is followed through, and, if the student subsequently doesn't turn up, this is followed up;
- uses contact with parents to support sanctions;
- makes contact with home (postcards, letters, phone calls) for praise as well as admonishment.

Three caricatures

In order to consider some aspects of classroom management that are perhaps particular to the Mathematics classroom, we will try to imagine three caricatures of teachers in their first few years of teaching.

The flapper

The flapper, let us call her Miss Jones, is the teacher who always seems in a panic and a rush. Students can sense your level of distress, as a wolf smells a vulnerable young lamb. You must be organised, especially as you set that tone in your classroom in your first few months in post. *The flapper* will bustle into the room just after the bell, with many students already lined up ready for the lesson. She is probably still carrying her mug of coffee, demonstrating that she hasn't really had time to sort herself out in the classroom.

As the class enters the room, Miss Jones is fussing with her register or the interactive whiteboard, trying to get the materials ready for the start of the lesson. Gradually, the class enters, and eventually ten questions are set to initiate the lesson. She makes these questions up as she goes along, and students write answers. Subsequently, she goes through the questions, but, come number 4, she has to take a moment to work out the correct answer herself. Students chatter while she works this out.

Eventually, the main body of the lesson starts, and she explains what the class is going to learn today. This includes a couple of examples, which she rummages to find in the book. She picks these examples on spec and, part way through, finds that in fact they have a complex twist that means that they are not the best examples to use to build student expertise. Finally, the class gets on with some consolidation, but unfortunately the textbooks are in another classroom. Miss Jones sends a student to get the books from Mr Smith along the corridor. This takes 3–4 minutes. In the meantime, the class is restless and has nothing to do.

Once the books arrive, it takes Miss Jones a moment or two to select the exercise needed and get the class going. After a few minutes, students want to check their answers. Miss Jones has not worked these out herself and takes time to do this before she can give feedback to the class. This disrupts the learning and means that students wait to move on, because they do not wish to attempt further questions until they know that they are correct.

Eventually, Miss Jones goes through the questions as the plenary in the lesson. This takes some time, and some students mark their own work, whereas others do not, daydreaming instead. Miss Jones picks on those students with hands up for answers, and a large proportion of the class is not involved. Just as the bell goes for the end of the lesson, Miss Jones realises that she meant to set homework. She can't quite decide whether to get them to get books back out and set it now or leave it to next lesson. She airs these thoughts with the class and eventually decides to leave it until next time.

The solution to the flapper

The solution is one word, *organisation.* Particularly in those early days, your organisation is vital. If you are disorganised, students immediately lose faith that you know what you are doing. You send signals that you do not fully understand what needs to be done and that you are not going to be the steady influence that they need to be reassured when they are struggling. Students want to know that you know what you are doing. If you come unprepared for the lesson and are at sixes and sevens throughout, you will never instil confidence in them.

As we have already discussed, one of the most important aspects to build in the Mathematics classroom is confidence. That starts with your confidence. So make sure lessons are fully planned in advance, that resources are all in the room ready for each lesson and that you watch the time to make sure that your plan for the lesson can be adapted and shaped according to how far the students have got.

The questions that you ask and explanations that you use must be fully planned. Make sure that the examples you use build skills gradually, and that each example adds another twist to the problem. For example, I ran a revision session last week for some students on solving linear equations. These were the examples that we worked through:

(a) $2x + 10 = 7$

(b) $5x - 7 - 2x = 12$

(c) $4x - 12 = 10 - 2x$

(d) $3(x + 13) = 50$

(e) $5(x + 4) - 2(x - 2) + 20$

(f) $\frac{3x + 5}{2} = 7$

(g) $\frac{3x}{2} + 5 = 7$

Can you identify what complication I am adding each time, and, hence, what teaching point I am making with each new question?

(a) A simple equation, where the focus is on understanding inverse operations and manipulation of the same thing on both sides of the equation, balancing.

(b) Collecting like terms before consideration of inverse operations and balancing.

(c) Collecting like terms, may include balancing if the terms are on opposite sides of the equals sign.

(d) Do we use inverse operations straight away, or would we multiply out the brackets?

(e) In order to collect like terms, we need to expand the brackets to this equation, but *watch out*. The negative before the second bracket throws in a common error where students simply put $-2x - 4$, without thinking about what will happen with that second term from the expansion and how this will be affected by the negative.

(f) Now, how do we deal with divisors? Do inverse operations work, and, if so, what happens when?

(g) Now the divisor is not the whole of the left-hand side but just one term. What difference does this make? How will we deal with the complication?

From a very simple topic, you can see the questions that I have chosen have been given careful consideration. They build gradually, helping students to build their confidence with the layers of complexity of the subject. Miss Jones needs to show she is organised in a generic sense, with timings, books and punctuality for the lesson, but she also needs to show that she is organised in her lesson delivery. For someone who has taught Mathematics for more than 20 years, this layering of questions comes relatively easily, but it does still require thought. For a new teacher, understanding of how to build up, or break down, skills so that you can help make explicit the pitfalls and foundations in each topic takes time and careful planning.

The dormouse

Miss Thomas will be our *dormouse*. As the class enters the room, she is seated at her desk, marking some books. She does lift her head, but does not acknowledge the students as they enter until nearly all the students are present. She has placed a task on every desk and she directs students to get on with this once they have arrived. She does not make explicit if students are allowed to talk as they do the task, and, once a hubbub of chatter has arisen, she tells the students that they are expected to complete the task in silence. She goes through the answer to the starter activity herself at the board and then explains the topic for the day.

As she does so, a student arrives late to the lesson. She asks him publicly why he is late. He goes into a lengthy explanation of where he has been and with whom. She spends 3 minutes listening to this (along with the rest of the class), and eventually he sits down. As the activity is explained, students ask questions throughout. Miss Thomas has not asked them to ask questions at this point, but she tackles each one as it arises, giving detailed and individual responses to that student, without engaging the rest of the class. This goes on for 10 minutes, until eventually the class is set to work. Miss Thomas gives the class some carefully prepared questions and sets them on their way. A student, Peter, asks if he can move to sit with someone else, Fred, as his usual neighbour is away. Miss Thomas enters into a discussion about how he doesn't normally work well with Fred and how last time this wasn't very productive. After many promises, Peter is allowed to move to sit with Fred.

Miss Thomas says that the students must complete ten questions. The class gasps at this and pleads that this is too many. Miss Thomas relents and says that they must do seven. The students work on the same questions, and, at the end of the lesson, Miss Thomas goes through the answers. She does so just as the bell goes. The students pack up without any instruction to do so, and she feebly tries to dismiss a row at time, only to have most of the class already gone.

The solution to the dormouse

The solution is again one word, *control*. Miss Thomas is not sending out signals that she is in control of what goes on in her classroom. She displays some aspects of organisation (the first, important step to control), but she does not then follow through her expectations and instructions with the class. She makes everything a negotiation, and students are 'empowered' to make their own decisions. Although, as you know, I am an advocate of independent learning and students taking responsibility for their work, this does not mean that they call the shots. The classroom is your domain, and, even when the tasks that you set are more open-ended, you are still in control of the choices that the students make.

I don't want you to interpret control to mean that we all need to be a forceful, imposing, rugby-playing physical presence in the classroom – far from it. One of the teachers I know who has the most control in the classroom has the quietest voice and the smallest frame, but is direct and clear about her expectations. As students enter the room, she acknowledges them, directs them to the task on the desk and is already setting an expectation of their work and concentration; other things do not distract her. The class has her full attention. If a student is late, it is acknowledged, and she makes it clear that a discussion will be had later in the lesson; for now, however, 'Sit down and get your work out because we have already started'. The conversation can be had at the end of the lesson or even in the lesson, in a quiet moment, discretely, but do not make it part of the public domain.

When Miss Thomas is completing her explanation, of course we wish for students to question and probe to ensure that they understand, but that process should be managed with confidence:

> I am going to explain this once through. Don't make any notes and to start with don't ask any questions. I will stop after I have got half-way to see if we have any elements that we don't understand. Be patient and be prepared to think about what is happening.

This way, our dormouse is being clear that questions will be good, but does not get sidetracked too soon and then lose momentum with the rest of the class. There is a time for questions, and other students are used to answer these, but she is in control of this.

Similarly, be clear about how students are expected to work. The dormouse is not explicit about when the class should be working in silence, in pairs or as groups. This is another aspect of control. Consider different phases of the lesson to allow for different interactions. I find a silent Mathematics classroom very dull, but there is a time for this, when students have to concentrate (really concentrate) on what they are doing, and there are other times when discussion with their peers will help their understanding. Be direct about what you expect – perhaps 5 minutes first in silence and then 5 minutes working quietly. In the starter activity, direct students to do it on their own first and then signal when they can share their response with their neighbour. You need to think about what the activity warrants and then make the class clear about what you expect. The best Mathematics classrooms have different periods in the lessons where there are different expectations, but the transition to each is directed by the teacher, not the students.

The same applies to the tone that Miss Thomas sets as the class enters the room. The students also need to know that they leave the room on her command and under her supervision. If you have a stray, don't run after them like an arm-waving lunatic, but make a mental note next lesson to tell that student (and the rest of the class) that they will go last for their misdemeanour. Your control requires follow-up action and a dogged sense that, if you ask a student to do something, you will follow it through until they have done so. This doesn't require you to bellow or shout; it just requires you to be persistent.

Don't think that I want your classroom to be a rigid environment where students are threatened and punished continually. We do not want Miss Thomas to spend all her time growling and seeming grumpy. We want her to build relationships with the students, through talking to them and engaging in conversation about their Mathematics and their wider schooling and interests. The critical thing is that she does this on her terms and when it is appropriate. To stop being a dormouse, Miss Thomas needs to demonstrate control through tough care, helping students to learn her routines and setting boundaries for what is expected. She needs to exert control in the classroom, but she should balance this with support and coaching.

The mate

Mr Wakeling is gregarious and funny. He can entertain the students and keep them engaged with his winning smile and his never-ending stream of banter. He has stories from his travels and he shares them constantly. He wants to be mates with the students, and it is really important to him that the students like him.

His classroom is fairly disorganised, but, rather than flap about this, he has a very laid-back approach, and, if something is not where he thought it was, he will 'fill' a few moments with his string of jokes while a student runs to collect the materials that he needs.

Mr Wakeling sets homework once a week, but he always marks it in class and very rarely takes the books in. When he does collect work, it is marked with a tick or a cross only, and he makes very little comment on the work that they have completed. Marking, to Mr Wakeling, is an unnecessary part of teaching, and he can't understand why some teachers give up so much time in the evening to provide such detailed feedback. In fact, he is almost resentful of those teachers who do this, because it makes him look bad.

The solution to the mate

Of course we want students to like us, and the positive relationship that is formed when they do like us is very powerful. However, students do not need or want you to be their friend, they want you to be their teacher. They will like you if they respect you and they will respect you if they consider that you are supporting their learning, don't ever give up on them and, ultimately, care if they achieve.

A sense of humour can be helpful in the classroom, and certainly allowing the students to see that you have one is vital. You mustn't be stern at all times, and the old adage, 'Don't smile until Christmas', will mean that you lose classes, as they sigh with a heavy heart every time they enter your room. Humour can make things memorable and create a warm dynamic in the classroom, but don't get carried away. You are there to teach them, not to entertain.

Behaviour issues specific to the Mathematics classroom

Ensuring good behaviour is the same in all classrooms, but there are some words of advice that I would give that are specific to Mathematics. You will encounter, if you haven't already, plenty of students (and sadly adults) who will say things such as, 'I am no good at Maths', 'How is this relevant to me?', 'What is the point in learning how to do this, I am never going to need this in a job', 'I can't do algebra' and, worst of all, 'I hate Maths'. Such comments will get in the way of setting a positive tone in the Mathematics classroom. You can enter into a debate about why Mathematics is worthwhile and how students might use a quadratic equation if they end up working in architecture and building bridges. You could start to espouse the virtues of the problem-solving, logical-thinking nature of Mathematics and how this will build transferable skills, whatever walk of life they enter into. Or you could spend time making sure that students are confident about what they are learning, have opportunities to explore and discover and gradually build skills so that they feel success.

In my experience, those who come with negative attitudes to Mathematics are those who have had a bad experience, perhaps have struggled with classes that have gone too fast and left them behind or have been boring, owing to lack of variety in the learning. If you consider all the aspects of planning and pedagogy that we have discussed and implement these in your classroom, you will go a long way to eradicating those negative attitudes.

One important aspect of building this confidence and avoiding disruption in your classroom is considering what strategies you develop for when students 'get stuck'. The Mathematics classroom is unique in that, if students face something they cannot do, it can be very immediate. When they write an essay in English, draw and label diagrams in Geography, examine sources in History or conduct and evaluate a practical in Science, there is usually something that they can say. In Mathematics, students can feel as if they have hit a wall and, hence, they down tools, stop working and start chatting, and then the rest is history (not medieval!). Set the tone in your classroom for how you want them to respond when they get stuck and perhaps display this on the wall and direct them to it when they want immediate attention from you in the lesson.

One often-quoted example is:

Book – Buddy – Boss

This implies that, before asking you how to unpick a problem, a student should first consult their book (this implies that they will have clear, worked examples to which they can refer). Then, they should discuss with those next to them how they might overcome the difficulty (this implies that there is a calm, but not silent, working environment that allows students to talk about what they are doing). Finally, they should ask you how to overcome the difficulty. Mathematics teachers can often be faced with students' need to depend on them for the answer. You might need to wean them off you with an approach like this, so that they have strategies for when they are stuck.

Sometimes, you will do challenging work that requires them to think, and they may get stuck. You need to learn to judge when to bring the class back together. When have they had enough time thinking it through, and when have you hit a critical mass of students demanding your attention, such that you cannot manage to get around to all of them to answer their questions? This judgement is a fine art. Step in too early, and they will not have been forced to think. Leave it too late, and the class may well have descended into mayhem, lost confidence and be a challenge to draw back together.

The teaching of Mathematics requires you to be able to tackle problems and unpick their misconceptions, so that they can continue next time without you. When you are looking at their working, ask them first, 'What do you understand?' and 'Where do you get stuck?'. In other words,

> Let us establish what you can do, what you know already that might help, and how you apply it to start the question. We can then find out where you have gone wrong and what you need to think about to address it.

In summary

The most critical lesson that I ever learned with regard to classroom management is that you have to be yourself. When I struggled during training with a very tough class, I tried to be like my school mentor. He was a 6-foot rugby player, whom the kids adored. If one of the lads gave him a hard time, he was likely to challenge him to a run next time he took rugby training. I tried to be larger than life and use a high volume for attention that he used when they got rather rowdy. Unsurprisingly, as a 22-year-old female, this didn't work! I learned that I have to use my own personality and style to direct how I manage the class; that the most important things are building relationships, caring about students and their progress, following things up like a dog with a bone and being myself. If you adopt all of these, you will be just fine.

ACTIVITY

Video-record yourself to analyse how you come across in the classroom. It may make you squirm, but it is the only way to see yourself as the students see you.

TALKING POINTS

- Are you using the micro-skills to set the tone in your classroom?
- How will you foster an environment that builds confidence rather than steals it?
- Do you fall into a parody such as our flapper, dormouse or mate?
- How will you build confidence in the classroom, through which variety of teaching techniques?
- How will you foster an environment where students have to think for themselves before they expect you to do it for them?
- How will you judge the moment when you need to step in and draw the class together?

Your thoughts

Note

1 D. Lemov (2010) *Teach Like a Champion: 49 techniques that put students on the path to college.* San Francisco, CA: Jossey-Bass.

10 Dealing with observations

This chapter will look at the difference between long-term progress and progress in a single lesson. How do you demonstrate that students are making progress over a short period of time, without resorting to interrupting learning to make false 'checks on progress'?

The game of lesson observations, quite rightly, is changing. Quite possibly by the time this book is in its fifth year, the game will have changed again. Therefore, I am going to avoid giving you tricks to jump through whatever hoops are out there at the moment, as you read this book. Good teaching to me is good teaching, and it isn't measured by 20 minutes in a classroom watching the teacher with the class. It comes from a much broader range of evidence than the observation itself. When watching a lesson, I always take the opportunity to talk to the students and look in their books and, when meeting with the teacher after the lesson, I always ask them to bring their markbook or assessment evidence with them for the class observed. That way, I can form a full picture of the progress those students are making and not base any judgement on whether the teacher can put on a 20-minute show when I have agreed to watch them.

So, despite what I said in the introduction to this chapter, I am not going to distinguish between long-term progress and progress in a single lesson. Progress is progress is progress! On the other hand, I wouldn't plan for my class do be doing a test in silence or a very complex introduction to a topic, if I was going to be observed, as I don't think that would allow me to demonstrate my teaching skills and how they help students to learn.

The following *essentials* are a set of prompts that my school uses to remind teachers about the ingredients of the best lessons. These are not all encompassing and are never used as a tick list for a lesson observation. They are merely a shared understanding of what outstanding teaching and learning mean and, hence, a guidance for teachers as they work. I might add that we have a very similar list of *essentials* for the students, to remind them of their responsibility for their own learning and how important it is that they think about how they learn and how they can be more effective as well (this is, after all, a job that involves them coming half-way!).

Nine essential elements of the best lessons

Evidence of progress

Do students know what they are expected to learn in this lesson (not just what they are going to do)? Do they know how it links to previous and future lessons? Are there appropriate expectations of progress for students who are SEN/FSM/G&T (special educational needs/free school meals/gifted and talented)? Does the teacher know who are vulnerable groups and show awareness of and plan for their needs? Is it clear to students what progress they are expected to make, and is this challenging? Is the progress evident and sustained, over the short and long term?

Standards

Are students seated according to a plan? Are they in correct uniform? Are coats off, bags on the floor, and log books on desks? Are they equipped for work? Have they been registered?

Sustained pace

Are there a variety of activities, and do the tasks maximise student progress? Does one task lead logically to the next? Does the teacher create clear transitions between tasks? Is the overall pace appropriate and purposeful? Are students given realistic deadlines and reminders of time?

Effective questioning

Are questions open or closed? Are they planned or spontaneous? Do they link to what students are learning, or are students being required to guess an answer in the teacher's mind? Is the teacher asking key questions in order to assess understanding and deepen learning? Do students get thinking time and oral rehearsal? Is there an alternative to putting hands up? Do students get to ask questions?

New learning

Are the students working harder than the teacher? Do students get to work independently at some point? Does the teacher avoid dominating the lesson through talking? Can students explain how what they are doing links to what they are expected to learn? Can they articulate what they should be able to do or to know by the end of the lesson? Can they show how it builds on prior learning? Does the end of the lesson review what has been learned? Can they make connections with skills and knowledge learned in other subjects?

Tackling behaviour

Are there high levels of engagement, courtesy, collaboration and cooperation? Is there an ethos that is calm, purposeful, secure and encouraging? Are expectations of behaviour made clear? Is praise genuine and purposeful? Are students given responsibility and independence? Is the first hint of off-task behaviour dealt with?

Intervention

Are interventions focused and timely, having a positive impact on the pace of learning? How are interventions with individuals or small groups organised? Are the right students selected? Do they benefit? Is the presence of teaching assistants maximised, or are there periods in the lesson when they are sitting listening to the teacher talking? What liaison takes places between teacher and teaching assistant, to ensure consistency? Does intervention help students to make progress?

Assessment

Are students able to demonstrate what they have learned, in writing and orally? Is understanding checked throughout the lesson? What kinds of question are asked, and how many? Is talk used to help students to articulate ideas and to deepen their understanding? Are adjustments made to the lesson where necessary? What responses are there to ongoing assessment?

Literacy and numeracy

Is the teacher taking responsibility for promoting high standards of literacy (reading, writing and oral communication) and numeracy, irrespective of the subject? Does the teacher draw attention to the key vocabulary of the subject? Are keywords displayed? Is the spelling of keywords highlighted and advice given on how to spell more accurately? Is spelling in books corrected? Does the teacher place high expectations on presentation of written work, irrespective of a child's background? Does the teacher model and demonstrate how to write in the subject, and how to read?

How do the essentials fit into my planning?

So, when planning for a lesson observation, I would consider how my teaching and the learning activities are realised in all these areas. I want the observer to see a class that is engaged, enjoying the subject, being made to think and, as a result, perhaps finding things hard. When they are faced with a challenge, however, do they have, and do I give them, the confidence to have a go at questions and to try to solve problems? All observations will come down to the progress that students make. If they are not faced with challenges, then they will not, ultimately, make progress. My job as a Mathematics teacher is to make sure that they can reason through problems, ask questions to fully understand a situation and grow in knowledge as a result.

Some basics: Whenever I am being observed, I will provide the observer with a lesson plan, so that they can see the context of what I am doing, and a seating plan, so that they can see where particular students are. The seating plan would contain names, any characteristics of students (SEND etc.), target level or grade and current working level or grade. This way, the observer can see which students are in the class and, if need be, talk to students who are making more or less progress, and it is all open from the start.

Be warned: don't try tricks and routines in your observed lesson that you don't normally do in any other lesson – the students will always trip you up. So the fact that, in an observed lesson, I will always have a task for the students to do as soon as they enter the room is not unusual. This will either be a task on the board, a common sheet on the table or, at my best, a personalised sheet on the table that asks the student a question that follows on from recent work and is based on my assessment of what they need to do to improve. This last one is cunning in that, immediately, I have shown that I am using assessment to inform my planning, and that this is then personalised and differentiated.

The lesson would always then involve some instruction, but I would be wary of talking for too long. I want to get the students active and participating in the learning straight away, and so the instruction will be minimal, and then they will get going on a task from which I want to draw conclusions. They might spend 5–10 minutes doing the task, which we then discuss, in pairs, in groups and then as a whole class. The 'think, pair, share' motto really works to show how you are building confidence and including all students in the learning.

After some conclusions drawn together, as a group, I will then set another task that consolidates what we have discovered, so that I can be sure that all have understood the learning.

It sounds so simple, doesn't it! The critical things are getting the students working, rather than you just talking, and making sure they are taking responsibility for that learning. If you go back to our discussion of mathematical pedagogy in Chapter 2, I concluded that the best mathematical learning included elements of exploration, explanation and consolidation. This happens over time, but it can also happen in a single lesson. If you are being observed and you are only tackling one element of this, then make sure that you are clear to the students (and, of course, your observer) how today's lesson fits into the sequence of lessons. So state it explicitly.

> Today we are going to explore a new area of Mathematics. I am going to give you a chance to try to discover the rule that we are then going to use. To do this you will need to think about what patterns there are, what conclusions you can draw and how you might use this.

This stating of the obvious brings the observer into the sequence of lessons, but it also makes the students aware of what they are doing and why. This explicit explanation of the purpose of the lesson is a practice that makes all feel involved and responsible for the outcome of the lesson.

An example: teaching Pythagoras

When you are being observed, be clear in your mind what the outcomes are. Let us imagine we are doing an exploratory lesson on Pythagoras. Students are cutting out squares drawn on the perpendicular sides of the triangle and are rearranging them on the square on the hypotenuse. This lesson is about visualisation and exploration, to ensure a deep understanding of the properties of right-angled triangles. At the end of a 50-minute lesson such as this, they are unlikely to be using $a^2 + b^2 = c^2$ in a 3D setting, but that might be the ultimate aim. Make your objective clear to the observer and the students. This might be set at the beginning of the lesson, simply as:

> Objective: to discover properties of right-angled triangles and consider how these might be applied.

But, at the end of the lesson, this might be linked to where it is then going:

> Objective: to discover properties of right-angled triangles and consider how these might be applied.
>
> Stage 1 To be able to use the rule $a^2 + b^2 = c^2$ in right-angled triangles.
>
> Stage 2 To be able to use the rule $a^2 + b^2 = c^2$ in complex diagrams that can be broken down into right-angled triangles.
>
> Stage 3 To be able to use $a^2 + b^2 = c^2$ in 3D shapes to find lengths.

In this way, once again, the students and the observer are involved in the journey and know what purpose the skills they learn today will have tomorrow.

In essence, the judgement of a lesson and the quality of teaching, for me, will come down to these seven things:

- Do you know your students?
- Do you know your subject?

Dealing with observations

- Do you have high expectations?
- Do you engage all learners?
- Do you plan for pace, variety and challenge?
- Do students understand what they are learning, and can they articulate it?
- Do you know when they are (and aren't) making progress? And what do you do to speed this up?

These questions would be what I would keep in mind when planning for an observation. If I am doing these all the time, then I shouldn't have to worry too greatly when I am being observed. The evidence will be there.

TALKING POINTS

- How will you be covering those nine essential skills in your observed lesson:
  - evidence of progress
  - standards
  - sustained pace
  - effective questioning
  - new learning
  - tackling behaviour
  - intervention
  - assessment
  - literacy and numeracy?
- How will you make sure that the students are working harder than you are, but allow time for delivery of a key concept or setting up an exploratory task where they discover the information for themselves?
- How will you make it clear that they are making progress (in the lesson and over time)? *No* tricks, remember. Make

this meaningful and appropriately placed. Don't let your 'check on progress' get in the way of progress.

Your thoughts

11 Applying for your first post

This chapter will consider what you should look for in that first school and how to give yourself the best chance of getting an interview. Chapter 12 will consider what to do once you get that interview.

What sort of school do I want to work in?

What kind of school do you wish to work in for that first post? You might be limited by geography as to where you apply, tied by living arrangements, family or friends. Whether you are restricted to an area, wish to work in an urban or rural environment or, quite simply, wish to be selective, you must think about the type of school in which you want to work.

For me, there are five school characteristics to consider:

- commitment to professional development;
- academic attainment;
- size;
- staffroom culture;
- mathematical pedagogy.

Commitment to professional development

This is the most important to me. You want to know if the school is one where everyone is learning, from student to teacher, from newly qualified teacher to experienced head of department. Is there a culture of reflection upon practice, and is there an ethos that we can all learn something, no matter what stage of our career we are at? If this pervades the school, you know that you will have an initial teaching experience that will help you to grow. People won't look at you oddly for asking how they would approach a certain problem, or if they have read a particular article in this week's *TES*.

How can you tell if this is the case? A visit to the school prior to your application might give you an opportunity to ask about the support for professional development, but I have to say I am in two minds about these preliminary visits. Although they give you a snapshot of the school, to avoid wasting an application, tours on such visits can be carefully plotted and give you answers that encourage an application, rather than tell you the full story about the school. Remember that, if you do ask to visit the school like this, this is the first point that you are 'on show'. As you get a feel for the school, it will be getting a feel for you, and arrogance, timidity, an overbearing personality or just being irritating might mean that you don't even get an interview. Personally, I like to sell myself on paper and make the interview day the first point at which they get to judge me in person.

The website is, of course, the first place to look to see what a school is like. How current is the site? Is there a feeling that this is regularly used to communicate with the outside world and present a positive image of the school. Can you download the school prospectus? What does it say about the school's commitment to staff development? Does the school have trainee teachers? Is it a member of any groups that share practice across schools?

Do you know of others who know the school? If it is local to your training provider, are there other trainees who have been placed there? What can they tell you about the school?

Academic attainment

I am fully committed to a comprehensive education and all that implies. However, you must consider the academic profile that will suit you for your first place of work. Look on the website or in the government performance tables to see how the school fares. Make sure you look, not only at raw attainment, but also at the progress that students make. This information is easy to access, and the beauty of Mathematics is that it will be there in detail for the whole school, but also for this department in isolation.

How confident is your subject knowledge? Do you want to work in a school that achieves very high academic attainment? If the answer is yes, then you need to consider that there will be pressure on you to know your subject well and be challenged by pupils who are aiming at the highest grades. Don't fall into the assumption that students will all be eager to learn and easy to teach where attainment is high. Students who attend these high-attaining schools can often think that they will achieve because they are at a good school and, hence, expect you to do all the work for them. Students may have good levels of prior attainment, but there will be an even greater pressure to get higher outcomes.

In lower-attaining schools, there will still be pressure to perform, but it will be different. Students in schools that are set in a more challenged environment are often more grateful for your expertise and support for their learning than those from more privileged backgrounds who expect it from you. I remember vividly the first few weeks that I worked in a challenging school, my second school in my teaching career, and students thanked me as they left the classroom. In my first school, this would have been a very rare occurrence. Although far more challenging in their attitudes, the students in my second school were overwhelmingly grateful when I showed that I cared for them and worked hard to help them to achieve.

Size

The size of a school can make the environment feel very different, and you will need to consider what will suit you best. In larger schools, you may well be in a department of up to a dozen Mathematics teachers. It will be a larger school physically, and the site may be quite spread out. Often in these schools, departments have their own work and social spaces, and this can create a very strong team ethos. Equally, they can feel removed from the school as a whole, and, particularly as a new teacher, the interaction with staff from other departments is useful. Will you grow in confidence more from being part of a small and nurturing team, or would you find this more suffocating and insular. This will vary, depending on the type of person you are, but is worthy of consideration.

Staffroom culture

The staffroom environment is something that is important to consider as you weigh up the type of school in which you wish to work. How staff work together and socialise together is worth considering, as you decide upon where you are going to apply. My own experience has been that, the more challenging the school environment, the more camaraderie between the staff. I have never been one for socialising at length with my colleagues, but the odd drink on a Friday night, with the chance to chat and get to know each other in a different environment, is important.

Do staff take opportunities to chat with different people in the staffroom? We are all creatures of habit, and I think that, without meaning to, we often sit in similar places and talk to the same people, but are staff flexible in this, and how do they chat to newcomers when they arrive? This will be difficult to ascertain unless you visit the school or are on interview. You might not want to create a circle of friends from your employment, but you should not isolate yourself, and how much staff interact with each other and in what manner are things to think about.

Mathematical pedagogy

I mention this last, not because it is the least important, far from it, but because it might be the hardest to ascertain prior to interview. You might get a flavour on a visit to the school, but this is unlikely to be conducted by a member of the Mathematics department, and, even if it is, they might not consider that the detail of this aspect of their work is relevant at this stage.

What do I mean by mathematical pedagogy? The first and most superficial aspect will concern groupings. Does the department put students in sets or bands? If so, when does it do this? Setting students for Mathematics is the norm for most schools soon after entry, but there are some who remain in mixed groups for the first year at high school, and I even know of some who are taught in mixed-ability groups all the way through until the GCSE examinations. If you cannot imagine how you would teach a class with Year 11 students ranging from A* to E, or would not wish to, then that will not be the school for you.

More subtly, how is Mathematics taught? Is it, as at one school I visited last week, taught at Key Stage 3 through a thematic exploratory approach, where the content evolves from the questions that students ask and how they approach the task? Or is it very content driven, with clear learning outcomes every week and term? Most schools, as I have advocated, achieve a happy medium that combines the two, but there are those that sit at the extreme ends of the spectrum. Which are you more comfortable with, and how can you discover the approach for a given school? The only way to find out the latter is to spend time on interview talking to members of the Mathematics department and asking about their approach. Do they have rigid schemes of work, and, if so, how do they work?

The application itself

Most application forms have the same standard elements of education and work history, disclosures, training and wider experience.

It is important that you include all details of your history to this point. From a child-protection point of view, a school will want to see what you have done throughout your academic and working life, and any blanks could look as though you have something to hide.

Remember to take time and care over your application. It seems so obvious to say don't make any mistakes, but I read applications all the time where people have spelled words incorrectly, have missed out key information or have clearly completed it in a hurry. The school wants to know that you care about this appointment from the word go, and a sloppily completed form might mean that you do not get in the door.

Other than accuracy, there is little influence that you can have over what you put on the nitty-gritty parts of the form. You can, however, think about what you have done that might be relevant or helpful to this application. Have you done any volunteering or working with young people? Do you belong to any educational organisations, especially with relation to your subject area? Are your hobbies outside school things that could be brought into school as an extracurricular activity? Schools are always looking for what you can bring beyond your classroom teaching. As someone new to the school, you might bring something new and different that will enhance its provision.

Your supporting statement or letter: three golden rules

The main area where you can influence the school's impression of you will, of course, be in your letter. I would say that you want to display three aspects of your development thus far:

1. What do you know of general teaching pedagogy, and how have you brought this into your classroom?
2. What is your philosophy for teaching, and how can you demonstrate your enthusiasm for your subject?

3 What concrete examples can you give of strategies that you have employed and that have been successful in your training?

I will give you an example of a letter of application. There is a danger that you might then copy this letter for your own. I will show you how this letter covers the three points that I mention, but please then make it your own. Never let your letter be more than two sides of A4 and make sure that you are clear and concise in your language.

ACTIVITY: THREE GOLDEN RULES OF YOUR LETTER OF APPLICATION

Below is an example of a letter of application. Take a highlighter and, considering my three golden rules, highlight examples of each rule within the letter (there will be many correct answers to this task of course, but mine is included later).

Highlight within the letter where I am tackling each of my three golden rules. Pick out keywords or parts of sentences, rather than full paragraphs. There may be more than one rule being tackled in each paragraph.

- Use yellow highlighter for 'general teaching pedagogy'.
- Use blue highlighter for 'my love for Mathematics'.
- Use green highlighter for 'Strategies I have employed'.

Dear Miss Upton

I should be grateful if you would accept my application for the post of Mathematics Teacher at Fictitious County High School.

I have been fortunate, during my time at Teacher Training School One, to benefit from an ethos that encourages all staff to reflect upon their practice; I have learned much from staff in the Mathematics department but also from those in other subject areas. I have taught classes across the age range and have developed not only my teaching pedagogy but also my subject knowledge.

I am committed to improving standards and encouraging all students to realise the important place that Mathematics has in society. In my own teaching, I see myself as someone with a willingness to 'step outside the norm' and I am enthusiastic about making Mathematics challenging and exciting for all students. Through a balance of exploration and discovery combined with consolidation and practice, I aim for all students that I teach to become confident users of mathematics, at a level appropriate for their goals and ambitions.

I belong to subject specific associations such as the Association for Teachers of Mathematics, as well as receiving regular updates from websites such as the National Centre for Teaching Excellence in Mathematics, in order to be informed of best practice in this critical subject for all young people today. I realise that some students do not find Mathematics easy and it is my role to ensure that they build confidence in their skills to be able to become confident users of both numeracy and mathematics in its purest sense in the future.

My experience has taught me that students respond to praise and that part of my role is to be interested in what they are doing and how they perform. In my teaching practice so far I have experimented with different forms of assessment and feedback. I have been assessing two things: which form of feedback are students more motivated by and what helps them to learn most. Unsurprisingly, given research of many before me, I have found that use of personalised, regular, meaningful feedback is best for both aspects of assessment. The use of grades is useful in order to demonstrate progress but it is essential that students can reflect on their learning and identify what they need to do to improve, not just know that today, they are achieving a grade C.

While teaching one particular group of Year 8 students, I have noticed their dependence upon the teacher for solutions, support and problem-solving. From this I have been developing strategies for students to work collaboratively to overcome challenges, before needing to 'ask the expert'. I subsequently used a structured approach while teaching the areas of shapes to see how I could make students more independent learners. Although it is hard in a short space of time to quantify something as complex as independence, students were certainly more prepared to try for themselves after a week of lessons.

I am keen in actively taking part in wider aspects of school life, as can be seen from my involvement in extracurricular activities during my teaching practice, particularly of a sporting nature, and would be keen for this to continue. I see it as an important, and enjoyable, way of gaining a wider insight into students' interests, and have found that it helps greatly to build relationships within the classroom.

I have a passionate belief in the power and elegance of Mathematics and the place that it holds in society and am committed to trying to foster this belief in my students. Although I see any teaching post as encompassing academic, pastoral and extracurricular, all of which I have experience in, Mathematics is my true interest. I would hope to improve upon the high standards already achieved by your school academically through an approach that would combine tradition and innovation.

Thank you for considering my application.

Here is my answer, where **bold** text represents green highlighting, italic text represents blue highlighting, and underlined text represents yellow highlighting:

Dear Miss Upton

I should be grateful if you would accept my application for the post of Mathematics Teacher at Fictitious County High School.

I have been fortunate, **during my time at Teacher Training School One**, to benefit from an ethos that **encourages all staff to reflect upon their practice**; I have learned much from staff in the Mathematics department but also from those in other subject areas. I have taught classes across the age range and have developed not only my teaching pedagogy but also my subject knowledge.

I am committed to improving standards and encouraging all students to realise the *important place that Mathematics has in society*. In my own teaching, I see myself as someone with a willingness to 'step outside the norm' and I am enthusiastic about making Mathematics challenging and exciting for all students. <u>Through a balance of exploration and discovery combined with consolidation and practice</u>, I aim for all students that I teach to become confident users of mathematics, at a level appropriate for their goals and ambitions.

I belong to subject specific associations such as the *Association for Teachers of Mathematics, as well as receiving regular updates from websites such as the National Centre for Teaching Excellence in Mathematics,* in order to be informed of best practice in this critical subject for all young people today. I realise that some students do not find Mathematics easy and it is my role to ensure that they build confidence in their skills to be able to become confident users of both numeracy and mathematics in its purest sense in the future.

My experience has taught me that students respond to praise and that part of my role is to be interested in what they are doing and how they perform. In my teaching practice so far I have experimented with **different forms of assessment and feedback**. I have been assessing two things: which form of feedback are students more motivated by and what helps them to learn most. Unsurprisingly, given research of many before me, I have found that use of **personalised, regular, meaningful feedback is best for both aspects of assessment**. The use of grades is useful in order to demonstrate progress but it is essential that students can reflect on their learning and identify what they need to do to improve, not just know that today, they are achieving a grade C.

While teaching one particular group of Year 8 students, I have noticed their dependence upon the teacher for **solutions, support and problem-solving**. From this I have been developing strategies for students to **work collaboratively** to overcome challenges, before needing to 'ask the expert'. I subsequently used a structured approach while teaching the areas of shapes to see how I could make students more independent learners. Although it is hard in a short space of time to quantify something as complex as independence, students were certainly more prepared to try for themselves after a week of lessons.

I am keen in actively taking part in wider aspects of school life, as can be seen from my involvement in extracurricular activities during my teaching practice, particularly of a sporting nature, and would be keen for this to continue. I see it as an important, and enjoyable, **way of gaining a wider insight into students' interests, and have found that it helps greatly to build relationships within the classroom**.

I have a *passionate belief in the power and elegance of Mathematics* and the place that it holds in society and am committed to trying to foster this belief in my students. Although I see any teaching post as encompassing academic, pastoral and extracurricular, all of which I have experience in, Mathematics is my true interest. I would hope to improve upon the high standards already achieved by your school academically through an approach that would combine tradition and innovation.

Thank you for considering my application.

Yours sincerely
W. J. Cooper

TALKING POINTS

- What type of school do you want to work in?
- What mathematical pedagogy do you prefer, and how will you find out how the Mathematics department operates?
- What can you put in your application form that will demonstrate your commitment to teaching and your experience with young people or in the area of Mathematics?
- What concrete examples of success can you talk about in your application? What more might there be to these to explain if you get to interview?

Your thoughts

12 Interview advice and likely questions

This chapter will be split into two: generic and Mathematics-specific interview advice. It will look at what the day might include, how to prepare and how to get a feel for the school, so that you make a genuine decision about whether the school is right for you. It will consider the interview itself (including if this is the most important part of the day) and what kinds of question you might face. It will give advice on how you can demonstrate your expertise and knowledge throughout the day.

Interviews have become more involved than when I started out in teaching. Interview candidates are now subjected to a range of activities, in order to best demonstrate their skills. Each part of the day plays its part in the forming of a judgement on the best candidate, and, even in the parts that seem 'less formal', be on your guard and consider how you are presenting yourself. It might sound like Big Brother, but the reality is you are being watched at all times.

The structure of the day

I will start with a structure for the day that is commonplace in most schools and will take you through, part by part, how you might tackle each aspect of the day.

Here is a structure to the day (with rough timings):

9.25	Arrival
9.30	Meet and greet with the head teacher
9.45	Tour of the school
10.30	Teach a lesson
11.15	Break with the department
11.30	Student panel
12.15	Informal chat with the head of department
1.00	Lunch with some staff
1.45	Interviews start

As you can see, you are likely to be kept very busy throughout the day, but there will be times where you can take stock and consider for yourself if this is the right school for you. I shall take each part of the day in turn and consider what you might gain from it and how you should approach it.

Before the day

You will obviously have applied for the job, having responded to an advert and considered what the school is like. You can get all manner of information about a school from its website, including the length of lessons, number of staff, academic results, extra-curricular activities on offer, exam board used for Mathematics and names of some key staff. Use this to your advantage. It can tell you all the nitty-gritty about the school and allow you to sound informed about a range of information throughout the day. You would look a bit of a wally these days, if you asked what time the school day starts and ends, if that was something you could have easily ascertained from the website.

Your arrival

You might wonder why I need to even mention this part, other than to say the blindingly obvious: don't be late! But this, for me, has always been a critical part of the day for me to see the school in action. Very often, your arrival time will be after the school has settled to work for the day. Some schools might invite you in early enough to attend a daily briefing and suffer the humiliation of the 'here are the candidates' introduction to a watchful group of staff. However, more common is a slightly later start, once the students have arrived, and the school has started its daily business. Use this time in the morning to see the school in action. Find out from the school website what time the school day actually starts and get there before this time. Seeing how students arrive at school can tell you a great deal about the catchment area of the school, the behaviour of the students and the expectations as they walk through the school gate. I often make my way to a side entrance, after parking my car, and watch the students arrive. Once, on a headship interview, I wandered into where the bike sheds were and bumped into the caretaker, who was clearing up the fag ends prior to the interviews for a new head teacher that day! Do the students come into school with their uniform in place? Are they putting out their cigarettes at the gate? Are the staff out on duty, and are they talking to the students?

Meet and greet with the head teacher

Depending on the format for the day, this might be the only time that you get to meet the head teacher, until the formal interview later in the afternoon. There won't be a great deal of opportunity for questions, and don't dive in with your list of twenty questions at this stage in the day. The day should give you a chance to find out information about the school, so don't expect to have asked all your questions by 10.00; you'll probably find that, by lunch time, even without you directly asking, they have all been answered.

This meet and greet might include other staff, such as the head of department, and is an introduction to the day, where you hear about the school in general and see the plan for the day. *Do* ask if you have any questions about the format of the day that you are presented with and, if you have made any requests for the lesson later (I'll come to this in more detail later on), make sure you check that they have been put in place. These situations are always slightly odd, as you are in the company of the other candidates. In these mixed situations, make sure that you are warm, friendly and approachable, not hogging the conversation, but equally not sitting meekly in the corner either.

Tour of the school

This might be with a member of the leadership team, or even with some students. You will probably end up feeling a bit lost; don't worry about where you are during the tour and making your way back: they aren't going to abandon you in the far corner of the school and set you on an orienteering mission to make it back to the staffroom. Use it to assess the tone of the school in a working day. If you are shown into classrooms (or even if you have to peer through the windows), are students on task? Are they working together or alone? If the teacher is talking, do they look like they are really listening, or are they chatting to each other or doodling in their books?

As well as looking into classrooms, look at what the fabric of the school is like. Of course, newer schools will have a shinier veneer than ones that were built back in the 1960s, but does it look as though the appearance of the school matters? Is there litter in the corridors? What is the quality of displays in corridors or key public places in the school? Are there pictures of students? Is there student artwork? What do the displays tell you about how much the school is proud of its students and what they do? A busy school will be subject to wear and tear, but does it look loved?

You will almost certainly be shown where the Mathematics classrooms are. Does each teacher have their own classroom? Are all the Mathematics classrooms in a cluster together? (It is really valuable, as a new teacher, to know that you will not be stuck out on a limb and will be near your head of department, should you need it.) Does the Mathematics department present itself as an area for learning?

Use the tour to ask more general questions with regard to other facilities that might be of interest to you. From a teaching perspective, what are the ICT facilities like? Does every classroom have an interactive whiteboard? Is there a suite of laptops or a classroom that you can use if you wish to use ICT with your classes? Asking a question about this will inform you, but will also set a flag in the mind of someone doing the tour that your lessons will be creative and varied, not just in rows in a classroom doing exercises 1–100!

Getting involved in an extracurricular club of some sort is really valuable in your early years (and beyond). You get to see students in a different light, and it gives you a different perspective on the school and what it stands for (does it care about developing the whole child?). When on the tour, ask what activities there are for students and show your interest in offering something. You will be finding out something about the school, but, at the same time, it will be finding out about you.

Break with the department

Don't be too nervous about this part. It is a less formal part of the day, but you want to come across as someone who can interact with other people and who will work well as part of a team. Don't be too put off if one person in the department is cynical or negative or a bit grumpy; one person does not a team make. Most importantly, take this time to chat to the staff (not the other candidates) and ask them about how long they have been at the school.

The student panel

This is often the most enjoyable part of the day. You have (hopefully) come into teaching because you like young people, and so this part should be enjoyable and should see you at your most comfortable. That said, beware – the questions that students come up with are perceptive and taxing. Here are just a few that they have come up with over the years:

- Are you strict?
- What is your favourite Mathematics topic to teach?
- If I struggle in your class, how will you help me to improve?
- What do you enjoy most about teaching Mathematics?
- How do you explain difficult concepts to students?
- How do you cater for students of different abilities in your class?
- If we asked students in your current school what they thought of you, what would they say?
- What will you offer the school other than teaching Mathematics?
- Is being a tutor important to you?
- How will you get to know the students in your tutor group?
- What do you read?
- And the daftest ever question that I have heard from a student panel and, I have to say, not a common type of question: If you were a type of cheese, what would you be?

Very often, the student panel will be conducted with a teacher observing. The teacher is looking at how you interact with the students, and how they respond to you. Don't be afraid to pose some questions back to them – what behaviour is like in school, for instance – but don't take over the questioning. Students are very perceptive about how you interact with them, and I have only disagreed with who is best for the job twice in all the appointments I have made. That doesn't suggest that this is the most influential part of the day, just that they make accurate judgements about whether you will be interested in them and will help them to learn.

Informal chat with the head of department

This might be as part of a group of candidates or, with a carousel of the other activities, on your own. Use this to find out what the Mathematics team is like, how the teaching is structured, and what support you will receive as a new teacher. As mentioned in Chapter 3, on planning, find out what sort of schemes of work the team has. Are staff expected to teach the same things at the same time? Do they teach in sets, and how often do these change? Do teachers all have a range of groups? As a new teacher, would you get all the bottom sets? Would you teach A level in your first year? How is this structured? Do teachers work in pairs and teach different modules, or do lessons run one after the other, from one teacher to the next?

Although I have run off a range of questions that you will want answers to, don't dive in with all these; the head of department will probably explain a great deal of this during their chat with you, and you will, hopefully, find many of your questions are answered throughout the day. Remember that you don't need to know every detail of how they do things, but you do want to ascertain what structures are in place across the team. The more there are, the more support you will likely have.

Be prepared to talk some Mathematics when you are with the head of department or other Mathematics teachers. Showing that you like this subject and want to talk about it is a good sign for any school. If the team members are worth joining, then they too will be happy to chat about the subject that they teach and, hopefully, are passionate about.

Lunchtime

There is likely to be some free time at lunch for you to wander the school. If this is not the case, take the time when others are being interviewed, and you are not, to get a feel for the place. Speak to other staff in the staffroom, but don't spend all your free time sitting, having coffee, and certainly don't spend it talking to the other candidates. You don't want to snub them, but, let's be honest, you are not going to be working with them, nor are you trying to impress them. Use your free time to wander around the school. How calm is the place at lunchtime? What activities are taking place? Are students allowed in their tutor rooms, or are they all outside, whatever the weather? This aspect is an interesting one if you consider that you want to spend your lunchtime marking quietly in your room. You won't get to do that if your tutor group is allowed in there as part of the school policy. If this is the case, is there a staff workroom? If so, is it packed at lunchtime?

What is the atmosphere in the dining room like? Are students polite to the catering staff, or do they treat them as second-class citizens. Students' behaviour towards staff at the school other than teachers can tell you a great deal about their respect and attitudes. Find the library. How busy is it? Are students working on computers, playing games or sitting reading? The library as a silent room doesn't tell you that this is a marvellous school, but it will tell you if students who are quiet, calm and perhaps a little quirky have a place to be themselves. Once, I had a student join, having moved into the area, who loved her new school because she could actually play at lunchtime. She was a Year 7 who was not very worldly wise and had previously been at a very small village primary school. Does the school give you the impression that the students can be themselves?

The day of interview is a two-way process, and you must make sure you find out about the school, as much as it finds out about you. The decision is yours as well as the school's, so make sure you feel informed, at the end of the day, about whether this is somewhere that you would like to work.

The lesson

I would say that, although an artificial situation, this is probably the most important part of the day. This, ultimately, is what you are going to do, so this is where you want to impress. I am going to talk about four key aspects of this show-stopper lesson. They are preparation, timing, activity and progress. In order to discuss each of these, I will imagine that you have been asked to teach factorising quadratic equations to a Year 10, set 3 group. You are told that they have been taught how to expand double brackets.

Preparation

Find out before the day what will be in the room that you are using. If you want to use a presentation, will there be facilities in the room for this? If you want the students to use compasses and rulers, will they all have them, or can there be a set provided so that every student has access to one? The last thing that you want is an activity where some students are not involved because they don't have the equipment that they need. If in doubt, and if the school has not been very forthcoming when you have enquired before the day, take it with you.

Ask about the class that you will be teaching. How many students will there be? Are there any SEN students? Will a teaching assistant be present? What level are they working at currently? This is critical to allow you to consider where to pitch the lesson. For the Year 10 class in our example, I would want to check what tier they are studying (foundation or higher), what grades they are expected to achieve, and how recently they were taught expanding brackets.

Make a lesson plan and take a copy with you to share with whomever is watching the lesson. This shows your preparation and allows you to share any reasoning for activities that you have chosen.

In this quadratics lesson, I would do an activity where the students have to match expanded and factorised expressions. Have these cut out already and prepared in envelopes for the students to use. Your classroom management of giving out and collecting such resources will also come through in the observation, so have these all prepared and easy to use.

Timing

First, with regard to timing, most schools will tell you a topic that they want you to teach, the group of students to whom you will be delivering the lesson and for how long. The length may vary from a 20-minute starter activity to a full 50-minute lesson. Whatever the length, plan it carefully. You should be able to judge how long your activities will take, and the worst thing would be to plan something that really takes an hour when you have been given 20 minutes. The figure shows my sample lesson plan for how I have broken down our 50-minute factorising lesson.

Notice that the teacher instruction is present, but it uses students to work through examples. I know that when you don't know the class this is risky, but using students at the board is a critical part of Mathematics teaching, and you want to demonstrate that this is part of your practice. Students are actively doing things for at least 30 minutes of the 50-minute lesson. During this time, wander around the room, guide and show how you interact with the students. You get a good idea of a teacher from how they check on students' working and consider how they will intervene to keep them on track. Think about the questions that you will ask to probe their learning.

Activity

What exactly are you going to get the students to do? How will you instruct them on what they need to do? I would always cover all bases with an observed lesson. You don't know the students – who might be SEN and who might be G&T – and so you don't know how good they will be at following spoken instructions. I would always do a presentation on the board of some kind, but I would accompany this with a hand-out with the key information and instructions as well.

Although, in a sequence of lessons, you might use a starter activity to pose a puzzle or a different problem, I would always use

Interview lesson plan

Subject: Mathematics	Teacher: J Upton	Date: 18/10/13	Lesson 3
Class: 10 set 3	No. of students: M F		Setted 3 out of 6

Learning objectives (students will learn to do):	**Context** (scheme of work/links to sequence of lessons):
Be able to factorise quadratic expressions with a range of different coefficients *ALL – factorise quadratics for all positive values* *MOST – factorise quadratics with a combination of positive and negative terms* *SOME – factorise quadratics with a coefficient of x^2 greater than 1*	*Find this out beforehand – what have they done and how recently?*

Differentiation (how have you catered for differing abilities in the lesson?)		
Lower ability Worksheet has a worked example for students to refer to, with reminders of factors of numbers and the key parts of the quadratic expression	**Middle ability** Consolidation is given on the worksheet so that students are able to embed the skills before moving on to more complex examples	**Upper ability** There are a range of questions, of varying difficulty for the students to tackle. This includes some stretch for those that understand the process of factorisation very quickly

Timings	Learning and teaching activities	Resources
5 mins	*Starter* *Factors of five numbers.* *Expand these five quadratic equations.*	*Sheet/board*
5 mins	*Go over the starter questions – what do we remember about expanding brackets?*	*Powerpoint*
10 mins	*Give the students a matching activity. This has cut out dominoes (15 in total) of expanded and factorised quadratics. The students have to match up the expanded and the factorised form to make a domino chain.*	*Sheet*
10 mins	*Go through the domino activity – display the correct solution on the board. Take four examples to then look at in more detail. Go through these with the class. Make clear the key features and the order in which the factorisation should be tackled.*	
10 mins	*Give out sheet with the explanation on clearly along with 15 questions for the students to tackle. The 15 questions have three groups. five all positive, five negative but manageable and five that are trickier. Judge how the group has responded to the initial tasks and their ability to guide where they start*	
10 mins	*Give the answers to the questions that the students have been doing. Ask them for their responses.* *Finally give them three questions to do in the last five minutes. Ask them to do two of the three questions. State that they are Easy, Medium and Hard difficulty. Mark these before you finish and collect them in.*	

the starter in an interview lesson to test the water with a skill that will be relevant to the lesson. It gives you something to come back to as part of the discussion and flow of the lesson and will aid you to judge the speed and ability of the group from the word go.

Will the activities that you have set promote discussion and enquiry? The reason for the domino activity is that it allows me to pose questions to the class, both as I wander and as we come together. What do you notice about the numbers in the expanded and factorised form? How could you work backwards from the expanded to the factorised? Which number would you look at first? How might our starter activity help you? Think about the questions that you will ask that will deepen learning and tease out understanding. Perhaps even put these on your lesson plan.

Progress

Our last aspect of the lesson is progress. Showing progress over such a short time frame can be very artificial, but you do need to demonstrate that it is important to the lesson. I would not do a show of hands of 'Who has learned something new today?', as I think this would not help learning in the long term and is not what you would normally do in the lesson. In our factorisation lesson, students have to complete two (from three) questions at the end. You can choose to mark these in the class or say that you would take these in to judge their success and progress. In the plan, I have suggested that you give the students three questions, but ask them to pick the two that they would like to try to demonstrate how much they have learned in the lesson. If you were teaching the group in the next lesson, this would give you an idea of what you need to do next, who to push on and who to consolidate with some easier forms.

Finally, when it comes to the lesson, be prepared to talk about how it went in the final interview. This is a critical part of their seeing you as a reflective practitioner. If the lesson was awful, say so; at least they will know that you know what is a good or a bad lesson. I have had one interviewee express complete surprise at this question and just say, 'I thought it was perfect'. They were immediately off the list as far as I was concerned.

The final interview

It used to be that schools asked all candidates to wait until the end of the day, after the interviews, and then, with you all together in a room, told you who was the lucky candidate. Thankfully, those days are over, and now, generally speaking, once your interview in the afternoon has been conducted, unless there are other activities around this, you are free to go. As mentioned earlier, don't miss this opportunity to see the school in action if you haven't had a full view already, before you go home.

So, let's consider the interview itself. You can expect this to be with three people, probably the head teacher, the head of department and another member of the leadership team. The length of the final interview will vary from one school to another, but expect anything from 30 minutes to an hour. The questions will vary and will probably depend on the area of expertise of the person asking. Expect three main areas:

1. your CV, previous experience and letter of application;
2. your Mathematics teaching;
3. your wider role in the school (and, hence, how you will contribute to it).

Remember the interview basics of making eye contact with all the panel, trying to keep hold of your nerves, giving yourself time to think before responding and sticking to the question asked.

Your CV, previous experience and letter of application

Expect them to go through your history, from education to experience in the workplace or as a volunteer. It is important that there are no gaps in your history, and they will more than likely use your application as a starting point to talk to you about what you have done before entering the teaching profession. If you have relevant experience in business or through work with young people, then make sure that you use this opportunity to tell them about things that you have done in greater depth. Interesting people make interesting colleagues, and the first part of an interview will nearly always be trying to settle you by talking about your experiences.

You might get the broad, non-teaching-specific questions such as:

- Why did you apply for this particular role?
- What are your core strengths?
- What are your weaknesses?

They are commonplace, but personally I don't think they tell me a great deal about your ability to do the job and tend to result in rather formulaic answers.

Your Mathematics teaching

Be prepared to show them why this job, and Mathematics in particular, interests you. These are a range of questions that might be posed to find out this aspect of you:

- Should every student have to do Mathematics?
- Why do we not have a more numeracy-based qualification for those who don't want to progress to A level and, quite frankly, only need to be able to use numbers proficiently? (Expect to get some questions where they are being provocative to get your view point. Be prepared to have a point of view on why your subject is valuable and what skills it develops, other than the content that it delivers.)
- Should we teach Mathematics in sets? How do we stop this from lowering the expectations for the least able? (You might

consider your answer to this in line with what the school does, but again be prepared to offer an opinion on what you think is best, and say why. You can always add the caveat that, of course, you have only experienced two schools and have not had the breadth of experience that others have to make this judgement.)

- Why do you think that the UK does so badly in the league tables for Mathematics education? (You don't need to be the next education minister to answer this question. Recognise that Mathematics teaching makes the news frequently, and that you should be aware of things that are being discussed.)
- What have you read recently that has interested you with regard to Mathematics teaching? What did it make you think? (In other words, do you actively take responsibility for your own professional development by reading resources aimed at the teaching of Mathematics?)
- How would you teach Pythagoras to a high-ability group of Year 9 students? How would you change that if it were a low-ability group?
- What is the best lesson you have ever taught? Why was it so good? (Think about progress, engagement and challenge as aspects to talk about.)
- How did you think the lesson this morning went? How did you know that the students made progress? What would you change if you did it again. (Don't go on about the room, not knowing the students' names etc. These are a given and don't show that you are being reflective on the actual learning that took place.)
- What is the worst lesson you have taught? Why did it not work? (Obviously, don't mention the lesson where they were climbing the walls and the head of department had to come and save you, but think about a topic that the students didn't understand and what you learned about how you explain, or a lesson where it was very teacher-led and you realised that you needed to make

the learning more active. This question can be an ideal way for you to explain what you think good Mathematics teaching is.)

- What do you know about the changes to the National Curriculum? (There is no need to be an expert on anything like this, but again be aware if there are changes afoot that will affect the teaching of Mathematics.)
- What does assessment for learning mean to you? What would this mean I would see, if I visited your lesson?
- How do you ensure all students make expected progress?
- Have you had any experience with teaching assistants in the classroom? How do you use them to help students make at least expected progress? (Think, again, about learning, not about classroom management. How would you communicate with a teaching assistant to inform them of what you are doing and how you want them to work with the students?)
- How do you motivate students who say that they hate Mathematics? (I want to know, not just that you like teaching the eager and the motivated, but that you gain as much pleasure from teaching those who have formed a negative opinion of the subject and whom you then turn around. Think about what you do in the classroom to motivate and engage – active learning, regular feedback and praise.)
- Imagine this scenario: You have been appointed and you have settled into the school. It is the third week of term, and students are starting to test you. You are giving an explanation at the board, and Jimmy starts talking to the person next to him. What do you do? (Scenario-based questions are always slightly artificial and, of course, they depend on the child, the group, the day and all manner of things. But consider how you deal with low-level disruption in the class. We all know that, if a student threw a chair through the window, we would call for help, but more important for the school to ascertain at interview is what your boundaries are for behaviour in the classroom and

how you would deal with this. Strike the balance between acknowledging the misdemeanour and dealing with it, but not allowing the student to hold court and sidetrack your lesson. Always look to deal with confrontation outside the public arena and stick to the adage, 'Praise in public, admonish in private'.)

Your wider role in school

Think about how you will contribute to the school. Schools are communities of people who need to work together for the common good of giving students the best possible educational opportunities. This means that, although 90 per cent of the time you will be in a class on your own, schools need people who work well with others and who give more to the school than the classes they teach. Also remember that you are likely to be a tutor to a group of students and responsible for their pastoral care, helping them to find their path in life and supporting them across all aspects of their education. This range of questions should give you an idea of the types of thing that you could be asked:

- Do you think that the role of the form tutor is important? Surely the most important thing is that you can teach Mathematics?
- What else would you like to get involved with in the school?
- What has your impression been of the school today? How does it feel similar or different to previous schools?
- How would you expect to work with other colleagues in the Mathematics department and in your year team?
- What would you do if a child started to cry in your lesson?
- What would you say to a child who asked to speak to you about something privately? (There is always likely to be some form of child-protection question. Remember that you can keep nothing private, protect yourself with open doors or others in the room, and inform a key member of staff of any interaction that you have with a child.)

Interview advice and likely questions

There will never be a cover-all set of questions, but think about how you can bring in examples of what you have done so far in questions that are asked, take time to give your responses and, if you can, try to enjoy the day. If it is the right school for you, and the interview brings the best out in you, you should enjoy it!

> **TALKING POINTS**
>
> - How will you use the interview as a chance for you to find something out about the school?
> - What case studies will you have in your mind that you might be able to apply to a range of questions? What skills will they show you have?
> - Remember that, in every interaction you have, you are on show. How do you want to come across?
> - What do you need to know about the school to be sure that it is right for you?
> - What else can you offer, other than your teaching of Mathematics?

Your thoughts

13 Your first term in post

This final chapter will take you through the trials and tribulations of that first term. It will include workload, setting the tone for learning in your classroom, how to seek support from your colleagues, planning for a full timetable and getting to know the routines of the school. It will consider how to set in place the foundation for a successful teaching career in these first months.

Managing it all

When we talked earlier about assessment and marking, I said that one of the most important aspects of a manageable teaching load is routine. This sets an expectation with your students and creates something that is manageable and expected for all. The very first task that I would set to when starting my first (and, indeed, any subsequent post) would be to look at the timings of my classes and how I might manage my weekly routine. I would not want to suggest that every week become a monotonous hamster wheel, but that knowing, generally speaking, what I am going to do and when will make my life much more manageable.

First, I'll be honest. Teaching is *not* a nine-to-five job, and, in my opinion (I will be shot down by some teachers), if it is then you are not doing a good job. You will not and should not manage all your preparation and marking in the 10 per cent (or more, if you are lucky) of time that you are not teaching. Expect to use evenings and weekends to manage this workload, but manage it you must. This workload does not mean that you are a poor, hard-done-by teacher who works harder than those in any other profession; it just means that this is the balance for the long holidays.

Planning your week

Let us imagine your week looks like this:

	Monday	*Tuesday*	*Wednesday*	*Thursday*	*Friday*
Lesson 1	7G		Yr 11 set 5	Yr 9 set 4	Yr 10 set 3
Lesson 2	Yr 10 set 3		Yr 8 set 6	7G	Yr 12 AS
Break					
Lesson 3	Yr 8 set 6	7G	Yr 9 set 4	Yr 10 set 3	Yr 11 set 5
Lesson 4	Yr 12 AS	Yr 11 set 5	Yr 10 set 3		Yr 8 set 6
Lunch					
Lesson 5	Yr 9 set 4	Yr 12 AS		Yr 11 set 5	

In our fictional school, Years 7–9 have three lessons a week, and Years 10 and 11 have four. Year 12 has five, but you only have three of their lessons. 7G are a mixed-ability group, until they are put into sets later in the year. There are seven sets in total. As a new teacher, you have five frees in the week.

The following are some questions to consider when planning your routines and structures for the timetable above.

ACTIVITY: HOW WOULD YOU PLAN YOUR WEEK WITH THIS TIMETABLE?

With your personal-life commitments – family, sport and leisure, household chores etc. – think about how you would plan what you would do and when in the week.

Consider:

- when you would plan the lessons;
- when you would set and mark homework;
- what times of the week are a no go owing to other commitments;
- if you work best at the weekend or in the evenings in the week.

My thoughts on planning

I would always plan a sequence of lessons at one time. Planning just one lesson at a time can be disjointed and does not make you consider the range of teaching strategies and activities that you are using over a longer period. I would plan the lessons for the week for a class in outline and then add further detail later if needed. I would do this with enough time to be able to ask others if they had any good suggestions and to plan any photocopying that needed to be done for the week ahead.

Planning ahead is also really helpful if you want to ask for help. Teaching is a very collaborative profession; despite the fact that you are mostly in a classroom on your own, other teachers are very helpful. However, if you want to ask advice on how to teach simultaneous equations, you can't do it the morning that you have that lesson. Think about when other colleagues are free (generally, in a Maths department, you will find a slot in the week where everyone is free together). This is a great time to be able to catch others and ask how they go about things. You will quickly decide who are your allies and who are those less willing to give you advice, but only by planning in advance will you give yourself the chance to use this valuable resource.

You have a really busy day on Monday, and so you don't want to be chasing your tail when you arrive on Monday morning. It all depends on what suits your pattern of work and home life, but would you stay on Friday to get your work prepared for Monday before you leave, or do you know that you will have a slot of time on Sunday afternoon to set out what you are doing in the coming week?

What evenings do you do other activities?

You don't need to give up everything else in your life when you start teaching. In fact, you will be a far poorer teacher and colleague, I would wager, if you do. But be realistic. If hockey training is on a Tuesday night, then, if you collect homework from Year 11 on Tuesday in the lesson and (as is right) want to get it back to them the next lesson, you either have to fit that in before you go, or when you return. If you are not someone who leads a happy life when you burn the midnight oil, then don't plan to collect too much work to be done on Tuesday evening.

When would you collect homework (assuming that you will set and mark this each week)?

Have a plan for the whole week. Yes, sometimes, when other things happen, this may need to have some flexibility. When you are in the midst of mock exams, have a long parents' evening and full reports to send home, you may not manage to keep to the same routine, but set out with this intention and, for your and your students' benefit, try to stick to it. From this timetable, I would do the following:

	Set	*Collect*	*Return*
Year 7	Monday	Thursday	Monday
Year 8	Wednesday	Friday	Monday
Year 9	Thursday	Monday	Wednesday
Year 10	Monday	Wednesday	Thursday
Year 11	Friday	Tuesday	Wednesday
Year 12	Friday	Tuesday	Friday

All groups have more than one day to complete the work set (allowing you to set the tone that they should come and ask you questions if they don't understand). This allows them to get into a good routine of, ideally, doing their work the night it has been set. Most groups have a couple of days for me to turn around the marking, allowing me a bit of flexibility if I should get asked out on a hot date unexpectedly! I have taken in one group's homework on the night of my activity, but this is my small set 5 in Year 11, and the marking demands for them are fewer than some other groups, by their size.

Another consideration is where you are teaching each lesson. Ideally, you will have a classroom base where you teach all your classes. This will allow you to have all the resources that you want to hand and to stay organised. If not, however, think how you can keep a stash of equipment in other rooms, so that you are not running back to your base to get equipment in the middle of the lesson, which makes you look disorganised to the students and will leave you in a flap.

It really doesn't matter how you set out your work, but make it work for you, your sleep patterns, your other life outside teaching and getting the work done.

What do I need to get right from the start?

Students want to learn

Remember that all students, underneath it all, want to learn and want to make a good impression. Even the most disengaged student would prefer to be praised for a job well done rather than be nagged for not getting it right or for not completing their work to the expected standard. When you start out (most likely at the start of a new school year, or at least a new term), they are eager to make the right impression and to do their best.

First impressions count

When I talk to trainee teachers, I often share with them a book that made me consider that first impression that you make with a class. The book, *Blink*,[1] by Malcolm Gladwell, examines a range of examples from the perspective of the first impression. The chapters vary from examples from sport and food production to the sixth sense of fire fighters. Of most interest to me, and I hope you as someone entering the teaching profession, is a chapter on the impression formed by students of their teachers in the first moments of meeting them.

The book focuses on a study with American university students and their lecturers. A survey is taken of student feedback on the quality of their lecturers at the end of the year. A similar survey is taken with different students, but the same lecturers, after a term. A video is used to then ask students their opinion after a much shorter period of time: first a few hours, then an hour, then a few minutes and, finally, a few seconds. The study found that the judgement made by the students who saw a video extract of a lecturer for just a few seconds was as accurate as the judgements made by those who had been taught by them for a whole year. Clearly, one needs to take a study such as this with a pinch of salt.

However, there are two important messages for anyone starting out in a teaching career:

1. First impressions count.
2. Students' judgements of who is and who isn't a good teacher are unerringly accurate.

Don't delude yourself: by the time that they start secondary school, students have been exposed to six years of teaching already. Whereas at primary school they may have only one teacher, once they start secondary school they probably have at least ten different teachers in a week, and they make comparisons. They have the luxury, which you don't, to see the same students, with the same classmates but with a different teacher, behave in a completely different manner from one room to another. They know very quickly which teachers to mess with and which you would be wise not to. Some of this is difficult as a new teacher. Reputations are formed, and knowledge of who is and isn't a good teacher is passed between generations and year groups in subliminal fashion. However, the impression that you make in those first few weeks is vitally important. We have covered some of this in consideration of our classroom behaviour in Chapter 7, but it is worth looking at again, as we consider those first impressions. Consider these ten things:

How tidy is your classroom?

I will mention display as a separate item, but how tidy are the chairs? Is the desk a sea of papers? Is there a coffee cup used earlier on the desk? Is the tone as soon as I walk in the room one of routine and high standards, or one of sloppy disorganisation? Get students to put chairs underneath the desks as they leave, so that the room is tidy for the next class. If there are any that have gone astray, then quickly straighten them before the next class comes in.

How hot and smelly is your classroom?

This might sound ridiculous and unnecessary, but consider the air of the room as your students enter. Sometimes, as you are in there all day, you can become oblivious to it, but there is nothing worse than walking into a room with a stale odour of teenagers' soggy jumpers after a wet lunchtime. Keep windows open and the air circulating. They might, every now and then, express concern that your room is cold as they enter, but it is amazing how quickly the place warms up when there are thirty teenage bodies in there. A hot, stuffy room is not conducive to concentration.

Welcome your class

Standing at the door, welcoming students as they enter the room, not just at the first lesson, but every lesson, sets the marker that they are entering your territory. I once read a book of comic anecdotes on teaching, and it suggested that you refer to everything in your classroom as belonging to you – 'my desks', 'my computer', 'my territory'. I wouldn't go this far, but let them know, with a warm and friendly welcome, that they have entered your space and time.

Language

Students respect teachers who are teachers; they don't want to be your mate, and you don't go up in their estimation if you use colloquial language to address them. 'Morning guys', 'All right Jack?', 'Hi there' and, even worse, a high five as they enter the room are all no go for me. Address them politely and with respect, but remember that you aren't trying to be their pal.

What is your classroom display like?

It takes time to get student work that can be displayed, but what are the first impressions of your students as they walk in the room?

Is it obvious that this is a classroom that they will learn Mathematics in? Keywords, common misconceptions and worked examples are easily created and put up in the summer holiday to set the tone as soon as they walk into the room.

What is there for them to do as soon as they walk in?

It doesn't take students long to get into the habits of your classroom. It is amazing how they remember your expectations and your routines and conform to them after a matter of weeks. However, these routines are much harder to establish half-way through the year, when you realise it has all gone a bit pear-shaped, than at the start. Demonstrate your expectation that this is a classroom where they come to work with a problem on the board as they enter. This might vary between a puzzle, some recap questions, a problem from the homework, a task on the desk or anything, but are they expected to be busy as soon as they enter the room?

Your confidence matters

I know that, in those first few weeks (especially as you face that first A level or Year 11 class), you will be petrified, but you must mask this. Students can smell the fear. Be confident in an assured, calm way. Lots of this will come down to thorough planning and prepared resources. You might also consider the aspect of body language and micro-skills that we talked about in Chapter 7, when considering good classroom behaviour management.

Get to know names

The use of names, as we mentioned in Chapter 7, is essential to good behaviour management. It will also set the tone for your expectations. The sooner you can learn students' names, the better. Have a seating plan and use it. It only needs a glance before you ask the class some questions for you to pick out a few key students. They will soon think that you know them all.

Marking and feedback

Practise what you preach when you set and collect work. If they see that you have given them thorough feedback and noticed when they have made an effort, they will respect you and know that, if they pay attention, they will learn with you. If you take in homework to mark, don't come to the next lesson without it, saying you left it in your car and got a lift in this morning. If you want them to meet deadlines, then you need to set an example. Students do lots of work in class time in Mathematics – you will build up different strategies for reviewing this, including going through answers in class as a first check. You must review these, however, when you collect their books. Identify what they are getting right, and wrong, and give them some feedback. Some teachers will do a cursory flick and tick, but the best will give students pointers as to where they have gone wrong in all work – classwork, homework and formal assessments.

Your appearance

We all know that first impressions can be shaped by appearance. Of course, the quality of a teacher's tie does not tell you if they are going to be able to teach you about surds, but a smart, businesslike appearance is important to set that tone. Once again, how can you tell them to tuck their shirt in and do their tie up, if you look a mess?

Get organised

Some of us are more organised creatures than others and do this naturally, rather than with explanation. You will find yourself awash with paper, emails and thoughts when you start out. It would serve you well if you were to organise these in some way.

For web-based resources, create some folders where you can start to store some ideas and resources. It can seem a good idea when you start to keep folders for each of the groups that you teach. I quickly found, one or two years in, that, when I was teaching a group at the same stage and on the same topic, but perhaps in a different school or a different year group, I didn't know where that Powerpoint display or worksheet was. So, perhaps using the curriculum areas of number, algebra, shape and space, and data handling, create a filing system for things that you create, use, steal and find. That way, when you are teaching that topic again at a later date, it should be easy to locate.

Many departments will have some form of shared resources like this within the team. This may include a filing cabinet with copied sets of resources that you can use. Make yourself familiar with these in your first term. That doesn't mean knowing their contents inside out and back to front, but find out what is there in a broad sense, so that you can look in their direction when you are looking for something particular.

You will be coming across new material every day in your first few years, so keep things organised. It will be a far lengthier job if you are sorting this out at a later date.

Mathematical developments

I would encourage anyone teaching to *read* about their subject. Journals, books, websites and, increasingly, blogs and instant media such as Twitter are all ways of keeping in touch with what is new and what others are doing in your subject. To put a list here would be worthless, as they change so frequently that it could be defunct in a short time. However, a few that have stayed constant (they have been around since I started teaching and have evolved successfully with changing media and journalism) and include articles and advice from all those at the forefront of Mathematics teaching are:

- NCETM: this is the National Centre for Excellence in the Teaching of Mathematics. With an online magazine, online

CPD, news and research stories, there is a wealth of material here to keep you engaged with what others are talking about mathematically.

- ATM: this is the Association of Teachers of Mathematics. ATM has always been a good hub for creative approaches to teaching and learning, a source of interesting activities for teachers and a powerful resource for research methods of what works within the classroom.
- NCTM: the National Council of Teachers of Mathematics is less practical in its resources than the previous two, but its *Journal for Research in Mathematics Education* is a worthwhile quarterly read.
- *Mathematical Spectrum*: also worthy of consideration are magazines and sites that are aimed at students. I have found many tasks and problems that I have used with my classes from this long-standing magazine.

The best way to gather resources and to avoid re-inventing the wheel is to look at what others have done. Personally, I never use a resource off the shelf; I always end up tweaking it in one way or another, but the starting point is invaluable.

A way to aid your planning: the three-student focus

Most schools will expect teachers to work collaboratively to support their professional development. One way of doing this in recent years has been something called lesson study. It is not an entirely new concept, just the development of a peer approach to professional development. The concept suggests that teachers should be grouped into threes. Once in these threes, each teacher considers an area that they would like to develop in their practice. They agree these and then they organise a time when the teachers will watch each other with a particular class. Two teachers watch the third and, after the lesson, they get together to discuss the learning in relation to the area of development that the teacher had set at the start of the exercise. This is very powerful research methodology and may be something that you come across in a school that you join. However, I do not mention it here for any of these reasons, but for one final aspect of the lesson study rationale. When considering what aspect of teaching they wish to develop, and which class they will observe to reflect upon this practice, the teacher also identifies three students in the class on whom they would like the other two teachers to focus.

Why do I mention this as an aspect worthy of consideration in your busy and overwhelming first term? I mention it as I think it is an incredibly powerful planning tool, rather like the lesson-planning sheet that we explored in Chapter 3. I would suggest that a manageable way to consider if you are setting the right level and type of work for your class is to consider three students. It will take you a few weeks to pick out whom you might have in mind, but, once identified, have in mind those three students when you plan your lessons. Of course, the trick to this is picking three students with differing characteristics. How you choose to identify that need not be by management-defined cohorts, but merely by how they manifest in your lessons.

So, a whole school-driven three-student focus might include:

- Peter: FSM student who has lack of support at home and low aspirations;
- Jane: SEN statemented student with Aspergers;
- Aluric: high-ability student who should achieve the highest levels and grades.

Instead, how about:

- Toby: very quiet student, doesn't like to offer answers in class, lets others dominate when working in groups, will answer questions when put under pressure to do so, but fears contributions in this manner;

Lesson plan

Subject: Mathematics	Teacher: J Upton	Date:	Lesson
Class:	No. of students: M F		Setted 3 out of 6

Learning objectives (students will learn to do):	**Context** (scheme of work/links to sequence of lessons):
Recognise and draw graphical regions for inequalities. We will be using a graphical tool to speed up the number of graphs that the students see in order to spend time thinking about the implications and conclusions rather than just drawing. *ALL – draw and recognise an inequality for a simple region e.g. $x > 2$* *MOST – draw and recognise simple non-horizontal regions* *SOME – start to combine regions with multiple inequalities*	*Students have discussed what inequality symbols mean and represented these on a number line. They have solved inequalities – using previously learnt techniques for solving equations. I have encouraged students to think about what the inequality means, not just applying a process.* *This is the starting point of what a graphical inequality looks like and the intention is that next lesson students will move to drawing complex three inequality regions.*

Student groups (who is … for ability relative to the group rather than the whole year)			
SEN	**Lower ability**	**Middle ability**	**Upper ability**

Timings	**Learning and teaching activities**	**Resources**
5 mins	*Starter – reinforcing and recapping mathematical language associated with topics learnt this term*	*Sheet*
1 min	*Recap what we have learnt so far about inequalities and the next step for today*	*Powerpoint*
5 mins	*Give students an introduction to the task for the lesson to include:* *How to use OMNIGRAPH (students have used once before)* *What the graph will show for an inequality* *How to draw a 'sketch'* *What is the task?* *What are the aims of the task?* *Tell students the pair they are going to work in (all complete a sheet but are expected to collaborate with their partner for support – pairs are ability chosen to push the most able and encourage their high shared dialogue)*	*Sheet* *Omnigraph* *Computers*
30 mins	*Pupils then move to the computer room (leave bags in room) take pen/highlighter and sheet to make notes on (given out)* *There are four stages of complexity of the graphs that students are asked to draw their graphs and write their conclusions on the sheet*	
10 mins	*Return to the classroom to share some conclusions and 'test' their theories with the mini whiteboards*	*Whiteboards* *Pens*

Homework
Sheet that asks students to draw three inequalities (of increasing difficulty) and then to try the challenge to draw the combined area that fits all three regions

Progress measure for each ability group:	**Support staff role:**
Upper – start to combine regions with multiple inequalities *Middle – draw and recognise simple non-horizontal regions* *Lower – draw and recognise an inequality for a simple region e.g. $x > 2$*	

- Mia: lacks confidence in her own ability, a competent mathematician but needs constant reassurance that she has got the right answer (which she very often has);
- Jesse: laid back (or lazy, depending on if you are giving him the benefit of the doubt) and does the bare minimum expected of him; enjoys Maths but hates showing his working and will often call out an answer without appreciation of others within the class.

Of course, consideration of the first three students has absolute merit and is necessary when you consider the progress made by everyone, but the second three may affect your planning in a more influential way.

- Have you planned a task that will allow Toby to discuss his solutions with others, to build his confidence?
- Does your planning show how students might check their own working and use others to know if they have reasonable responses?
- Have you been explicit in your methods of the level of detail in presentation required, and how are you going to enforce that all students (especially Jesse) do this?

Within the lesson-study model, the observing teachers then focus on these three students when they watch the lesson and consider how the task, lesson structure and teaching help them to make progress. Thinking about three students not only focuses your planning, but also your reflection. It can help to consider what you need to do next time, and how much longer you need to give each topic. It should help you to think about the range of activities and teaching methods that you are using over a longer period of time and if, through the use of each of them, any one of your three focus students is developing more or less than the others.

School routines

Finally, for this chapter on surviving your first term in post, get to know your school and teachers in other departments. The staffroom ambience is different in each school, and often in larger schools you will find that teachers in different teams rarely meet if they have their own department offices and work rooms. If the opportunity to meet other teachers is not as easy as in a busy staffroom every day, then make time to talk and interact with other staff. When you are on duty, talk to other teachers, offer to take a club with someone else, talk to tutors of students that you teach and get to know others in the school.

Schools are places of routines and structures; get to know what they are. There will be common expectations in your Mathematics department. When I was head of department, I ran the following things to ensure consistency across the team: a twice-weekly faculty detention (only for use when the teacher had already applied their own sanction and it had failed), letters home at half-termly intervals (I collected lists from teachers for different standard letters that were then posted home; these included problems such as no homework and lack of effort in class, but also praise and acknowledgement of improvement), coursework catch-up club, revision sessions that were open to all students etc. These types of structure and routine can be really helpful in enforcing your expectations.

Alongside routines in a department, also take time to get to know the expectations on staff at other times. Are there morning briefings? Are you expected to go to assembly with your tutor group? Who do you see if you are worried about a student? Are there forms for pupil misbehaviour? If so, to whom do they go? Getting to know the routines and patterns of a school is important to settling in and helping you to find your feet in that first term.

TALKING POINTS

- How are you going to use those first few weeks to set the tone of your expectations?
- What are your expectations? Are they high enough?
- What classroom routines are you going to have that will help you to manage behaviour, workload and expectations?
- How can you keep on top of learning on the job and wider mathematical developments?
- How will you keep a work–life balance?

Your thoughts

Note

1 M. Gladwell (2007) *Blink: The power of thinking without thinking*, p. 206. New York: Back Bay Books.

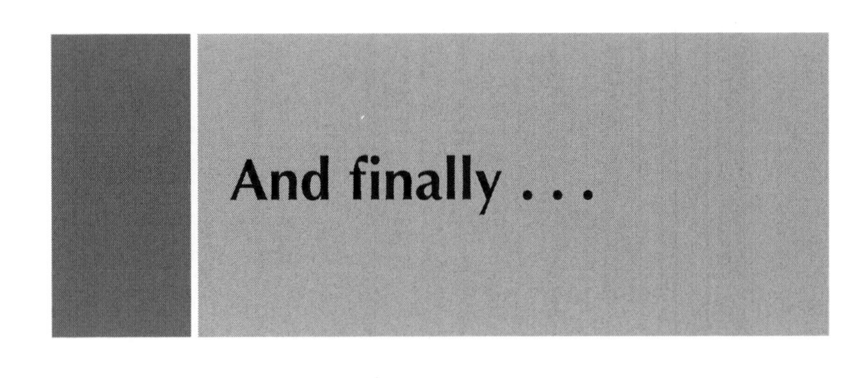

And finally . . .

Here are ten things that I wish I had known at the start of my teaching career:

10 Not every student will love Mathematics, or your style of teaching.

9 It won't always be right to stick to your lesson plan.

8 You are not trying to be their friend, you are trying to be their teacher.

7 There is much to learn in your first year, but it will all get easier as time goes on.

6 You will never stop learning about how to teach, and young people will always surprise you.

5 Parents can be awkward, but generally, when they are, it is because they care about their child and, although they might be difficult to handle, they are better than the ones who never take the time.

4 Lessons need to be varied, and sometimes this means quiet consolidation while you sit down.

3 Marking can pile up – get into a routine and try to stick to it to make it manageable.

2 You might cry some days, but there will be more when you laugh and smile.

1 You will make mistakes – but learn from them and share them to resolve them.

I hope that, from pedagogy to curriculum, from discipline to marking, this book will have helped you to set your first foot into teaching with confidence and a spring in your step. Enjoy the job; I have always loved it and always will.

Appendix

Here is my response for the activity in Chapter 2.

Lower ability

Recognising congruence (triangles that are the same)
Recognising types of triangle (isosceles, equilateral, scalene)
Measuring angles (using a protractor)
Calculating angles at the centre of the triangle using 360° knowledge
Calculating two other angles in the triangle using 180° triangle angle knowledge and properties of isosceles triangles
Conjecture – what if there are more/fewer points on the edge of the circle
Patterns and generalisation – forming rules in words/algebra to generalise

Higher ability

Calculating angles at the centre of the triangle using 360° knowledge
Calculating two other angles in the triangle using 180° triangle angle knowledge and properties of isosceles triangles
Conjecture – what if there are more/fewer points on the edge of the circle
Patterns and generalisation – forming rules in words/algebra to generalise
Angles in other shapes with their points on the edge of the circle – cyclic quadrilateral
Angles in a circle – angle at the centre
Influence of the radius on the categorisation of the triangle and the implications for right-angled triangle trigonometry
Area of a triangle from non-right-angled rule
Sectors – arc lengths, chords, sectors and segments

Bibliography

Beadle, P. (2010) *How to Teach*. Carmarthen, Wales: Crown House.

Black, P. J., Harrison, C., Lee, C., Marshall, B., and William, D. (2002) *Working Inside the Black Box: Assessment for learning in the classroom*. King's College London Department of Education and Professional Studies.

Bloom, B. S., Engelhart, M. D., Furst, E. J., Hill, W. H., and Krathwohl, D. R. (1956) *Taxonomy of Educational Objectives: The classification of educational goals. Handbook I: Cognitive domain*. New York: David McKay.

Boaler, J. (2010) *The Elephant in the Classroom: Helping children learn and love maths*. London: Souvenir Press.

Chaiklin, S. (2003) 'The Zone of Proximal Development in Vygotsky's analysis of learning and instruction'. In Kozulin, A., Gindis, B., Ageyev, V. and Miller, S. (eds) *Vygotsky's Educational Theory and Practice in Cultural Context*, pp. 39–64. Cambridge, UK: Cambridge University.

Collins, Mr. 'Mathematical Concepts Wall (for want of a better name).' See: http://mrcollinsmaths.blogspot.co.uk/2013/01/mathematical-concepts-wall-for-want-of.html?m=1

Cowley, S. (2010) *Getting the Buggers to Behave*. London: Continuum.

Department for Education, National Curriculum for Mathematics. See: www.gov.uk/government/publications/national-curriculum-in-england-mathematics-programmes-of-study/national-curriculum-in-england-mathematics-programmes-of-study

Duhigg, C. (2012) *The Power of Habit: Why we do what we do in life and business*. Toronto, Canada: Doubleday.

Eastaway, R. and Wyndham, J. (1998) *Why Do Buses Come in Threes? The hidden mathematics of everyday life*. London: Robson.

Bibliography

Eastaway, R. and Wyndham, J. (2002) *How Long is a Piece of String? More hidden mathematics of everyday life*. London: Robson.

Gilbert, I. (2007) *The Little Book of Thunks: 260 questions to make your brain go ouch!* (Independent Thinking Series). Carmarthen, Wales: Crown House.

Ginnis, P. (2005) *The Teacher's Toolkit: Promoting variety, engagement, and motivation in the classroom*. Carmarthen, Wales: Crown House.

Gladwell, M. (2007) *Blink: The power of thinking without thinking*. New York: Back Bay Books.

Great Maths Teaching Ideas. 'You've never seen the GCSE Maths curriculum like this before . . .' See: www.greatmathsteachingideas.com/2014/01/05/youve-never-seen-the-gcse-maths-curriculum-like-this-before/

Hodgen, J. and Wiliam, D. (2005) *Mathematics Inside the Black Box*. London: NFER-Nelson.

Kahneman, D. (2012) *Thinking, Fast and Slow*. London: Penguin.

Lemov, D. (2010) *Teach Like a Champion: 49 techniques that put students on the path to college*. San Francisco, CA: Jossey-Bass.

Lim, C. S. (2007) 'Characteristics of Mathematics teaching in Shanghai, China: Through the lens of a Malaysian', *Mathematics Education Research Journal*, 19, 1, 77–89.

Maslow, A. H. (1943) 'A theory of human motivation', *Psychological Review*, 50, 370–96.

Mathematics Enhancement Programme. See: www.cimt.plymouth.ac.uk/projects/mep/

Paulos, J. A., *Innumeracy: Mathematical illiteracy and its consequences*, New York: Hill & Wang, 1988.

Quigley, A. 'Hunting English.' See: www.huntingenglish.com

Ryan, W. (2011) *Inspirational Teachers, Inspirational Learners: A book of hope for creativity and the curriculum in the twenty-first century*. Carmarthen, Wales: Crown House.

Singh, S. (1999) *The Code Book: The science of secrecy from Ancient Egypt to quantum cryptography*. New York: Anchor.

Singh, S. (2013) *The Simpsons and Their Mathematical Secrets*. London: Bloomsbury.

Smith, J. and Gilbert, I. (2010) *The Lazy Teacher's Handbook: How your students learn more when you teach less* (Independent Thinking Series). Carmarthen, Wales: Crown House.

Vygotsky, L. S. (1978) *Mind in Society: The development of higher psychological processes*. Cambridge, MA: Harvard University Press.

Wilkinson, R. G. and Pickett, K. (2010) *The Spirit Level: Why equality is better for everyone*, London: Penguin.

Willis, J. (2010) *Learning to Love Math: Teaching strategies that change student attitudes and get results.* Alexandria, VA: ASCD.

Index

Index